THE IMPACT OF CLIMATE CHANGE

THE WORLD'S GREATEST CHALLENGE IN THE TWENTY-FIRST CENTURY

THE IMPACT OF CLIMATE CHANGE

THE WORLD'S GREATEST CHALLENGE IN THE TWENTY-FIRST CENTURY

CAROLYN FRY

NEW HOLLAND

Published in 2008 by New Holland Publishers (UK) Ltd
London • Cape Town • Sydney • Auckland

www.newhollandpublishers.com

Garfield House, 86–88 Edgware Road, London W2 2EA,
United Kingdom

80 McKenzie Street, Cape Town 8001, South Africa

Unit 1, 66 Gibbes Street, Chatswood, NSW 2067, Australia

218 Lake Road, Northcote, Auckland, New Zealand

10 9 8 7 6 5 4 3 2 1

ISBN 978 1 84773 116 6

Publishing Manager: Rosemary Wilkinson
Editor: Giselle Osborne
Design: Alan Marshall, Heron Recreations
Illustrator: William Smuts
Cartographer: John Plumer
Production: Melanie Dowland

Origination by Pica Digital PTE Ltd, Singapore
Printed and bound in Malaysia by Times Offset (M) Sdn Bhd

Photographs:
Front cover: Large ice flows © Corbis. Spine: Polar bear © Corbis. Back cover: Line of wind
turbines © Corbis.
Page 5: Air pollution from industrial chimneys © Corbis.
Page 7: Top: Polar Bear © Corbis; Bottom: a flooded street in the West End suburb of Gloucester,
UK, in 2007 © Corbis.

Contents

Introduction 8

CHAPTER ONE
Our changing climate 12

CHAPTER TWO
How the world works 52

CHAPTER THREE
Predicting the outcomes of climate change 94

CHAPTER FOUR
Mitigating the impacts of global warming 128

CHAPTER FIVE
Taking responsibility for our actions 165

Earth history timeline 184
Bibliography 192
Sundry documents 200
Index 204
Acknowledgements 208

Introduction

I began writing this book in September 2006 and completed it in February 2007. This is a short period of time in which to write any 80,000-word text, but it was a particular challenge given the complexity and fast-changing nature of the topic. The struggle I have faced to keep abreast of the changes happening to our planet over just six months is a good illustration of how urgently we need to tackle global warming. When I started writing, the latest analysis of data (published March 2006) from the World Meteorological Organization Global Atmosphere Watch (WMO-GAW) Global Greenhouse Gas Monitoring Network showed that, by 2004, the concentration of carbon dioxide (CO_2) in the atmosphere had reached 377.1 parts per million (ppm), a new high. By November 2006, when the figures for 2005 were announced, the figure had crept up to 379.1ppm. Similarly, when I began Chapter 1, the latest figure for the extent of global warming from the Intergovernmental Panel on Climate Change (IPCC) was for a 0.6°C rise between the late nineteenth or early twentieth century and 2000. As I typed away at Chapter 4, the panel's latest figures had revised the figure up to 0.76°C for the period to 2005.

In the intervening months, climate scientists on either side of the Atlantic pronounced that the UK and USA had experienced their hottest years since records began. Australia and China also sweltered as they experienced extreme droughts. Australia's was the worst on record for many regions; China's affected 18 million people. Meanwhile, India, southern Africa, Niger, Ethiopia, Alaska and the north-west USA all suffered severe flooding. The deluge that fell on the Greater Horn of Africa in October and November 2006 brought the worst flooding seen for half a century. In India's case, the flooding had begun in May, with the unusually early arrival of the monsoon season, and continued into August. And sea-level rise, caused by the expansion of warmer water in the oceans, also began to take its toll. As 2006 came to a close, Lohachara, a small island in the Bay of Bengal, disappeared beneath the waves, the first inhabited island to surrender to sea-level rise caused by global warming.

In February 2007, the IPCC confirmed what many scientists had believed for a long time: that Earth's rising temperature and associated climatic anomalies are the result of human activities. The ever-increasing number of factories, refineries, power stations, cars and aeroplanes that have been spewing gases into the atmosphere since the industrial revolution are to blame, because they have enhanced the natural greenhouse effect. This is the process by which the Earth regulates its own temperature. Essentially, the planet's surface absorbs light energy from the sun and then re-radiates it out as heat energy, which is trapped by greenhouse gases

such as carbon dioxide and methane in the atmosphere. If there were no greenhouse effect at all, the planet's temperature would be some 33°C cooler, similar to the moon, which has no atmosphere. Equally, raising the concentrations of greenhouse gases in the atmosphere starts the planet on the road to overheating. And that is just what we have been doing since the industrial revolution.

Although the world has only just woken up to the fact that human-induced climate change is upon us, scientists have been aware for a century that it could happen. As far back as 1896, Swedish scientist Svante Arrhenius predicted that the CO_2 that was being added to the atmosphere by the factories springing up across Europe could raise the Earth's temperature; he suggested that coal-burning would raise the concentration of the gas by 50 per cent in 3,000 years. In fact, it has increased some 35 per cent in a little more than a century, and is still rising by 2 parts per million (ppm) per year. Even if we had put no more greenhouse gases into the air after 2000, we would still be in line for a temperature rise of 0.1°C per year because of the length of time greenhouse gases remain in the atmosphere.

Sea level rise caused by melting ice and the expansion of warmer water is becoming a real threat for many coastal communities.

As more and more people move to cities, our demand for fuel and energy is rising

Carbon dioxide can linger for a hundred years; therefore pollution emitted in Arrhenius's lifetime is still potentially warming the planet today. Scientists have warned that, even if we manage to keep CO_2 levels to no more than 550ppm, which may be a struggle, we could still experience a global temperature rise of 2°C, 3°C or even 4°C by the end of the century. This could melt the Greenland ice sheet, bring droughts to crop-growing regions, cause severe flooding in coastal cities, and expose millions more people to malaria and other diseases.

Our only hope if we are to avoid dangerous climate change is to make deep cuts in greenhouse gases in the very near future. The political tool in place to enforce cuts is the Kyoto Protocol of the United Nations Framework Convention on Climate Change. This aims for the world to make cuts of five per cent on 1990 levels by 2008–2012, but it is having little effect, not helped by the fact that the world's largest emitter, the USA, has not ratified the deal. In early 2007, several reports highlighted the poor progress being made by individual nations towards cutting emissions. One announced that the US Government had missed all 34 of the deadlines set by Congress for requiring energy-efficiency standards. Another declared that UK plans to cut CO_2 were doomed because its climate change policy relies too heavily on voluntary reductions. Scientists remain hopeful that cuts can be made through a combination of increased efficiency, the use of new renewable technologies, and technological innovations enabling the capture and storage of carbon away from the atmosphere. However, having taken a century to decide that human-induced climate change is for real, we may have as few as ten years left to make the required changes before we commit ourselves to dangerous climatic shifts.

This book aims to explain how climate change has emerged as the most urgent challenge ever to face humanity. Chapter 1 looks at the scientific journey of discovery that led scientists to realize that humans were capable of controlling climate; it also explains the basic science of the greenhouse effect and the contributing gases. Chapter 2 explains how the world works geographically, and provides a digest of the latest published scientific findings showing how its natural systems are already changing. Chapter 3 explains how computer modelling can be useful in predicting future climatic changes and offers an insight into the shifts we can expect in different parts of the globe as the planet's temperature continues to rise. Chapter 4 assesses the action taken so far at international and regional levels and looks at what options we have for switching to a low-carbon global economy. And Chapter 5 considers what actions we can all take to reduce our own carbon footprint. Many uncertainties remain as to how the world will respond to the elevated levels of greenhouse gases that we are continuing to inflict on it. But it is clear that, unless we stop using fossil fuels to power the planet, we shall all be living in a very different world in a matter of decades.

Carolyn Fry

Our changing climate

HOW WE LEARNED THE WORLD IS WARMING

There is no doubt the world is warming up. Scientists have been measuring air and sea-surface temperatures at points around the globe since the mid-nineteenth century. Today, several thousand stations record the air temperature over the land, while automated buoys and volunteers onboard ships measure the warmth of the sea surface, and satellites take the Earth's temperature from space.

The University of East Anglia (UEA), US National Aeronautics and Space Administration (NASA) and US National Oceanic and Atmospheric Administration (NOAA) each run temperature-monitoring programmes. These centres independently collate data, make mathematical corrections to ensure that local characteristics, such as variations in altitude, do not influence the figures, and present yearly analyses of their findings. Thanks

People cool off in the fountains at Paris Trocadero during the summer of 2005.

to their efforts we know that, apart from a levelling-out of the temperature between the mid-1940s and mid-1970s (when sun-blocking particles from volcanoes caused a slight global cooling), the world's temperature has been rising steadily since around 1910. The latest figures tell us that the planet is now 0.76°C warmer than it was in the latter half of the twentieth century.

According to NASA Goddard Institute for Space Studies, five of the warmest years were between 1997 and 2007. In descending order they are: 2005, 1998, 2002, 2003 and 2006. The other temperature-taking groups agree that these years have been the warmest, although the order varies a little according to how they carry out their analyses. In July 2006, parts of the Low Countries, northern Germany and Poland recorded average temperatures that were between 5°C and 6°C above average. In the UK, the Met Office confirmed July 2006 as the hottest month experienced since records began in 1914. And China's State Flood Control and Drought Relief Headquarters reported that drought had affected 11 million hectares across western, central and north-eastern parts of the country. Around Chongqing, where the drought was the worst since 1951, the blazing heat dried up two-thirds of the streams and rivers, 471 reservoirs and 10,000 wells. And in January 2007, as this book was being completed, the month was shaping up to be the second-warmest on record in the UK and the hottest in the Netherlands for 300 years.

If you do not know very much about the Earth's climatic trends, you may think that such rather frightening statistics could be explained by natural causes. After all, geologists know that, in the past, the Earth experienced both warmer and colder periods than today. You can read more about this on page 184. For example, climatic clues contained in sediment cores taken from lakes in southern Italy show that Europe was much warmer 130,000 years ago; there were even hippopotamus wallowing in the Thames at that time. However, we know from research conducted in the early twentieth century that the majority of climatic shifts that have affected Earth over the past few millions of years occurred naturally as the result of variations in the Earth's orbit around the sun. Because scientists now understand the timing of these movements very well, they know that the recent global temperature rise cannot be attributed to them. Another ice age prompted by orbital shifts is not due for many thousands of years; some researchers believe human-made warming could prevent Earth from ever experiencing one again.

Cold beginnings

The scientific journey of discovery that led to this understanding began in the late eighteenth and early nineteenth centuries, when geologists were trying to explain how deep scratches found on rocks in the Swiss Alps came to be there. In 1837, the president of the Swiss Society of Natural Sciences in Neuchâtel, Louis Agassiz, announced that he thought the marks were the work of ice. Because the gouges were often observed in rocks far from the modern-day glaciers, he concluded that, at some point in the past, Earth had been covered in ice and snow to a far greater extent than was the case at that time. His announcement shocked the religious scientific world, who considered that the boulders and sediments strewn around the mountains had been deposited in the catastrophic flood on which Noah and his paired creature companions had floated to safety.

It was fortunate for Agassiz that he made his announcement in the same year as Queen Victoria's coronation. As the industrial revolution gathered momentum across Europe, and the rise of empire stoked demand for information on foreign terrains and the mineral resources they harboured, governments began funding expeditions to the farthest reaches of the planet. These yielded such a plethora of evidence suggesting ice had once covered much of northern Europe and large parts of North America that, by the mid-1860s, the idea was accepted on both sides of the Atlantic. The American palaeontologist Timothy Conrad noted, "M. Agassiz attributes the polished surfaces of the rocks in Switzerland to the agency of ice, and the diluvial scratches, as they have been termed, to sand and pebbles which moving bodies of ice carried in their restless course. In the same manner I would account for the polished surfaces of rocks in Western New York".

With the theory accepted, scientists wanted to know what had caused the ice to expand and, more importantly, whether such an "ice age" would return. The discovery of glacial sediments interspersed by peat that could only have formed in warm conditions made the search all the more urgent; clearly the climate was capable of switching between hot and cold states. During the nineteenth century two scientists, the Frenchman Joseph Alphonse Adhémar and Scotsman James Croll, both tried to prove that variations in the Earth's path around the sun had in the past caused the planet to fluctuate between warm and icy conditions. However, lacking mathematical training, they were unable to state their case definitively. That was left to Milutin Milankovitch, a Professor of Applied Mathematics at the University of Belgrade, who made it his life's aim to prove, once and for all, that astronomical changes were the driving force behind ice ages.

Planetary trigger

Thanks to the combined efforts of Adhémar, Croll, Milankovitch and earlier astronomers such as Kepler and Newton, we now know that the Earth's path around the sun affects the global climate in the following ways. Firstly, the tilt of the Earth's axis, at around 23.5° from the vertical, produces seasonal changes in our weather. This is because, as the Earth follows its elliptical path around the sun, the top of the axis is at times tilted towards the sun and at times away from it. When the top of the axis points away from the sun, the northern hemisphere experiences winter while the southern hemisphere enjoys summer. Six months later, when the Earth has

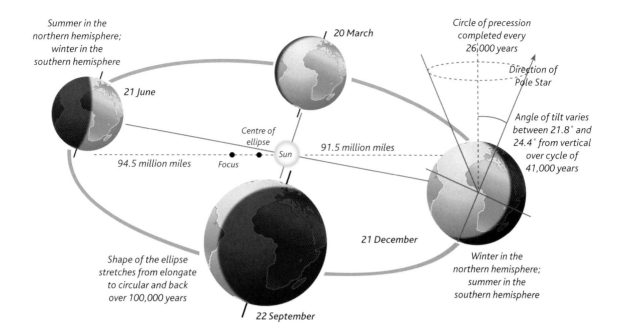

Summer in the northern hemisphere; winter in the southern hemisphere

21 June

20 March

Circle of precession completed every 26,000 years

Direction of Pole Star

Centre of ellipse

Sun

91.5 million miles

Angle of tilt varies between 21.8° and 24.4° from vertical over cycle of 41,000 years

94.5 million miles

Focus

21 December

Shape of the ellipse stretches from elongate to circular and back over 100,000 years

Winter in the northern hemisphere; summer in the southern hemisphere

22 September

traversed half the way around its orbit, the reverse is the case. However, as the Earth spins, the axis nods up and down between 21.8° (less tilted) to 24.4° (more tilted) from the vertical over a cycle of roughly 41,000 years. This varying tilt changes the amount of heat received in the polar regions by altering the angle at which the sun hits the Earth's surface there during summer. A greater tilt shifts the Arctic Circle (which marks the latitude at which the midnight sun and polar winter night begin) to a lower latitude.

Rather than being circular, the Earth's orbital path around the sun is elliptical. The sun is located not at the centre of the ellipse but at one focus; the other is empty. This means that, as the Earth makes its annual journey, it is sometimes closer to its glowing energy source and sometimes further away. It is at its closest on around 3 January each year, when it is 91.5 million miles from the sun. Six months later, on around 4 July, it is at its most distant position, three million miles further out in the solar system. The seasons change as the Earth passes through cardinal points on or near 21 December, 20 March, 21 June and 22 September. December 21 marks the point during the year at which the North Pole leans to its greatest extent away from the sun; this is the shortest day and marks midwinter in the northern hemisphere. In the southern hemisphere this date is midsummer's day.

As the Earth spins on its axis, the gravitational forces of the sun and the moon pull at it, causing it to wobble a little. This means the top of the axis describes a circle in space that completes a cycle approximately every 26,000 years. This causes the "precession of the equinoxes", essentially the slow advance of the four cardinal points around the elliptical orbit. The gravitational influence of the other planets in the solar system in turn causes the Earth's elliptical orbit to rotate independently in a counter-

Cyclical variations in the way Earth moves around the sun have triggered major climate changes in the past.

clockwise direction in the same plane. The combined action of the two is called the general precession. The result is that the shifting of the equinoxes completes a cycle every 22,000 years. Today, winter begins in the northern hemisphere when the Earth is close to the sun, near one end of the ellipse. But 11,000 years ago, winter began when the Earth was much further from the sun at the opposite end of the ellipse.

One further astronomical cycle concerns the shape of the Earth's orbital path. The shifting interplay of gravitational forces in the solar system causes the ellipse to stretch from circular to elliptical and back over roughly 100,000 years. This is termed the eccentricity of the orbit. Eleven thousand years ago, the Earth's orbit was very elongate, or eccentric, while for the past 10,000 years it has been more circular, or less eccentric. When the orbit is circular, provided the sun's energy output stays constant, the Earth receives the same amount of heat every day of the year. When it is elliptical, on some days the Earth is closer to the sun than on others. It receives more heat when it is close to the sun and less heat when further away, but the total amount of heat received through the year stays the same. The effect of the changing eccentricity, combined with the general precession and nodding angle of tilt, means that the Earth receives changing amounts of solar radiation seasonally and over cycles of 100,000, 41,000 and 22,000 years.

By studying the chemical composition of foraminifera shells, scientists have built up a picture of past retreats and advances of ice sheets.

Having calculated that the geometry of Earth's orbit changes according to these very long cycles, Milankovitch wanted to know how much solar radiation struck the Earth's surface during each season and at each latitude. He felt that ice ages might begin or end when astronomical changes in solar radiation caused ice sheets at certain latitudes to expand or contract. After plotting graphs showing the changing levels of radiation hitting the Earth at eight latitudes between 5° North and 75° North, he discovered that a decrease in axial tilt caused a drop in the radiation received during summer, while a decrease in the Earth-Sun distance during any season caused an increase in radiation in that season. The strength of the effects varied with latitude; the influence of the 41,000-year tilt cycle was strong at the poles, but less important in the tropics. Conversely, the 22,000-year precession cycle was small at the poles and large at the equator.

He deduced that, when orbital positioning reduced summer solar radiation at northerly latitudes to a critical level, the ice no longer melted during the warmest season and as a result ice sheets gradually began to advance, ushering in a new ice age. When the situation changed, this heralded the dawn of a new warm, inter-glacial phase. When Milankovitch compared his radiation curves with contemporary geologists' plots of ice-age timings derived from geological rock sequences, there seemed to be quite a good match. However, in the absence of accurate dating methods, there was no way to tell for sure.

Missing pieces

One of the problems of studying terrestrial geology is that wind, rain and frost cause rocks to break down into their mineral constituents, which are then washed far away by streams and rivers. As a result, rock sequences are often incomplete and it is not easy for geologists to deduce how much material has disappeared. From the late nineteenth century on, geologists sought to overcome the problems of studying interrupted terrestrial sequences by extracting sediment cores from the bottom of oceans and lakes. Layers of sediment constantly build up as the microscopic plants and animals that live in the brine die and fall to the sea bed. The scientists surmised that if they could extract long enough cores they would be able to obtain complete layered sequences dating back through past ice ages and interglacials.

Over the years, the geologists became adept at extracting cores and interpreting the data they contained. Specifically, they were able to glean information about global ice cover and temperature from the chalky shells of microscopic organisms called foraminifera contained within the cores. Seawater and the air we breathe contain two types, or isotopes, of oxygen. Isotopes are different atoms with the same chemical behaviour but with different masses. Atoms are made up of protons, neutrons and electrons. The number of protons sets what the element is, but the number of neutrons and electrons can vary. Oxygen has eight protons and usually contains eight neutrons to form ^{16}O, but in some cases there are nine or 10 neutrons. These give rise to ^{17}O and ^{18}O isotopes.

The ^{18}O isotope is heavier than the ^{16}O one, which means that some molecules of seawater (H_2O) are heavier than others. When water is evaporated from the oceans, the lighter molecules with fewer neutrons preferentially evaporate to the heavier ones. This means the vapour is enriched in lighter molecules and depleted in heavier ones. When the temperature of the air mass falls, prompting condensation, the heavier molecules condense out first. So the rain or snow that falls is initially enriched in the heavier molecules. If the temperature continues to fall, the condensation will contain decreasing concentrations of the heavy molecules. Ice sheets are particularly depleted in ^{18}O, so during ice ages, when large volumes of water become locked up in ice sheets, the oceans become progressively enriched in heavier oxygen molecules. The varying ratios of ^{18}O and ^{16}O present in sediment layers of deep-sea cores directly reflect the advances and retreats in global ice cover over the millennia.

As ocean-core science advanced, so did the researchers' ability to accurately pinpoint geological events in time. In the 1940s, scientists developed the technique of radiocarbon dating, which enabled them to place geological events in time as far back as 40,000 years. Then, in the 1960s, geologists discovered that they could date rock sequences according to their magnetic "signature". The Earth's magnetic field has at times been weaker or stronger than it is today and has sometimes reversed direction altogether. The magnetic character of a particular time is locked into sediments and rocks as they form, thus providing geologists with a means of dating rocks to a particular magnetic time over millions of years. By matching the changing sequence of magnetic signals present in terrestrial rocks to those found in deep-sea cores, and fixing dates at the top of cores using radiocarbon dating, they were able to date layers reasonably accurately.

The evidence that finally proved Milankovitch's astronomical theory for ice ages came in 1976. Jim Hays of the Lamont-Doherty Geological Observatory (now the Lamont-Doherty Earth Observatory), Columbia University, New York, USA, John Imbrie from Brown University, USA, and Nick Shackleton of the University of Cambridge, UK, combined magnetic dating with oxygen-isotope analysis to show fluctuations in ice cover represented in core sediments dating back half a million years. Their findings showed that the rhythmic advance and retreat of ice back through the centuries fluctuated according to three cycles of approximately 100,000, 41,000 and 22,000 years.

As Milankovitch had suspected, the varying amounts of solar radiation reaching the Earth's surface as it spun and wobbled its way around the sun had plunged the planet into repeated ice ages for at least a few hundred thousand years. Most importantly, the findings showed that the "normal" state of the Earth during this time had been full ice age. Interglacials had been rare and short-lived. Subsequent scientific studies have revealed that these orbital changes have a bearing on climate cycles not just at high

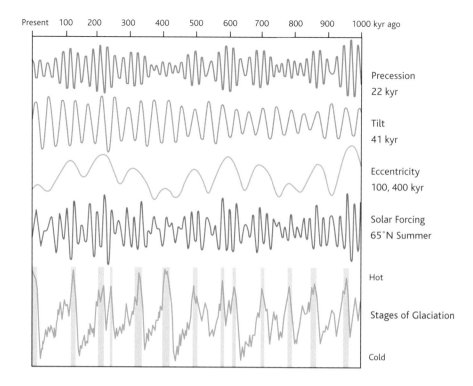

Present 100 200 300 400 500 600 700 800 900 1000 kyr ago

Precession
22 kyr

Tilt
41 kyr

Eccentricity
100, 400 kyr

Solar Forcing
65°N Summer

Hot

Stages of Glaciation

Cold

Graph showing how natural cycles have initiated repeated ice ages and interglacials in the past.

northern latitudes but across the whole world. For example, we now know that changes in the Earth's orbit and angle of tilt starting 9,000 years ago ultimately weakened the African monsoon and dried out the once-green Sahara desert.

Natural amplifiers

Although the evidence for astronomical forcing of ice ages was now very convincing, scientists were puzzled as to how small changes in solar radiation could have such an extreme effect on the Earth's climate. They began to wonder whether there might also be some natural feedback mechanisms at work. These might act to promote ice formation once the astronomical trigger had initiated a decrease in temperature at particular latitudes. One suggestion was that the sparkling white surface of existing ice sheets might reflect solar radiation back, causing the already low temperature to drop further and thus encouraging the ice and snow cover to slowly advance. However, layers of lake sediments preserved in rocks dating back some two hundred million years show that the Milankovitch cycles were also influential at times when Earth was warmer than today and free of ice. Such a feedback might work in a world that already had ice, but the proposed ice sheet feedback mechanism could not be responsible for coaxing temperatures into freefall if there was no ice in the first place.

In the early decades of the nineteenth century, when Joseph Adhémar first began thinking that solar radiation might influence ice ages, another scientist, John Tyndall, was tackling the conundrum from a different perspective. Trained as a physicist, he had become interested in Agassiz's

Svante Arrhenius, who predicted that CO_2 added to the atmosphere by human activities would raise the Earth's temperature.

ice-age theory and the motion of glaciers after holidaying in the Alps. He felt that changes in the composition of the atmosphere might be the driving force behind ice ages. He set about finding out by developing the first "ratio spectrophotometer", an instrument capable of measuring the abilities of gases such as water vapour, carbon dioxide, ozone and nitrogen to absorb heat.

His experiments revealed that the cocktail of "perfectly colourless and invisible gases and vapours" that make up the atmosphere exhibited vastly different abilities to absorb and transmit radiation. He found that oxygen and nitrogen, the main constituents, absorbed very little but water vapour, carbon dioxide and ozone were very effective absorbers of radiation energy at certain wavelengths, even when present only in small quantities. He concluded that of the constituents making up the atmosphere, water vapour is the strongest absorber of radiation and therefore the most important gas controlling the Earth's surface air temperature. He said that without water vapour the Earth's surface would be "held fast in the iron grip of frost". He also considered CO_2 to be an important influence in controlling the temperature of the Earth. In a paper entitled "On Radiation Through the Earth's Atmosphere", published in 1863 in *The London, Edinburgh and Dublin Philosophical Magazine and Journal of Science*, he likened CO_2 to a blanket that trapped heat in the same manner as glass in a greenhouse.

Thirty-three years later, and a decade after Karl Benz had patented the very first gasoline-powered motorcar in 1886, Swedish scientist Svante Arrhenius presented a paper to the Stockholm Physical Society entitled *On the Influence of Carbonic Acid* [carbon dioxide] *in the Air upon the Temperature of the Ground*. This article considered what impacts differing levels of CO_2 and water vapour might have on the surface temperature of the Earth. In making his presentation, Arrhenius drew together observations made by other eminent scientists including John Tyndall. He argued that fluctuations in trace constituents of the atmosphere, namely carbon dioxide, could greatly influence the heat budget of the Earth.

Like Tyndall, Arrhenius's interest lay in understanding what role CO_2 might have in initiating or perpetuating ice ages. He calculated that

CO_2 levels would need to sink to between 0.62 and 0.55 of their late-nineteenth century concentration in order to lower the temperature between latitudes 40 and 50 by the four or five degrees required to induce an ice age. But he also noted that the "temperature of the Arctic regions would rise about 8° or 9° Celsius, if the carbonic acid increased 2.5 to 3 times its present value". Some years later, Arrhenius published the book *Worlds in the Making*, in which he described the "hot-house theory" of the atmosphere. In it he stated that the Earth's temperature is about 30°C warmer than it would be due to the "heat-protection action of gases contained in the atmosphere". His calculations indicated, correctly, that if the atmosphere had no carbon dioxide, the surface temperature of the Earth would fall about 21°C, and that this cooler atmosphere would contain less water vapour, resulting in an additional temperature decrease of approximately 10°C.

By 1904, Arrhenius had turned his attention from studying how reduced levels of CO_2 might instigate ice ages and become interested in what impact CO_2 emissions from the new industries springing up in European cities might have on the atmosphere. He noted that, "the slight percent-age of carbonic acid in the atmosphere may, by the advances of industry, be changed to a noticeable degree in the course of a few centuries". He thought that an increase in atmospheric carbon dioxide due to the burning of fossil fuels could be beneficial, making the Earth's climates "more equable", stimulating plant growth and providing more food for a larger population. However, although Arrhenius was correct in assuming that factory emissions of CO_2 could warm up the Earth, his prediction that coal-burning would take 3,000 years to double CO_2 levels was way off. In fact, carbon dioxide concentrations have risen by 35 per cent in little over a century.

Rising CO_2

In 1938, nine years after Arrhenius had died, English engineer Guy Callendar spoke out in support of the Swedish scientist's hothouse theory. Piecing together temperature measurements from the nineteenth century, he noted an appreciable rise. When he measured CO_2 levels over the same period he discovered that they had also risen by around 10 per cent in 100 years. The existence of an increasing greenhouse effect was hotly debated until post-war funding yielded new studies and data. In 1956, Gilbert Plass confirmed that adding CO_2 to the atmosphere would increase the amount of long-wave infrared radiation (heat) absorbed. He suggested that industrialization would raise the Earth's temperature by just over 1°C per century. When Charles Keeling spent two years measuring CO_2 lev-els in the atmosphere in Antarctica and above the Mauna Loa Volcano in Hawaii from 1958, he discovered that even in that short space of time the concentration of CO_2 had risen. The graph of CO_2 levels recorded at Mauna Loa since Keeling's initial studies shows a steady rise to the present day.

When Arrhenius had first predicted that CO_2 might drive up temperature, his findings were given scant attention; as Milankovitch's astronomical theory gained support, the idea that atmospheric composition might influence ice ages was forgotten. But then, in the 1980s, researchers at the University of Bern in Switzerland reported their analysis of carbon dioxide bubbles in cores taken from the snow of the polar ice caps. Just as layers of sediments taken from the sea floor had provided scientists with data on fluctuations in ice cover, so ice cores now revealed changes in climatic composition.

The scientists had realized that when snow falls, air is trapped between the flakes. As more snow piles up on top, the layers become squeezed to form ice, but some air remains trapped in bubbles. The scientists measured the amount of carbon dioxide in the trapped air bubbles and found that, 20,000 years ago, when the last ice age was at its coldest, the amount of carbon dioxide in the air was about 180 parts per million (ppm) compared with 280ppm in the early nineteenth century before widespread fossil-fuel burning began. Around 16,000 years ago, when geological evidence suggests the ice sheets began to melt, the level of carbon dioxide in the atmosphere rose until it reached roughly the same concentration as that just before the Industrial Revolution. The two rival ideas for the cause of past ice ages were now both implicated in driving climate change.

Subsequent analysis of CO_2 levels in three Antarctic ice cores has shown that concentration of the gas starts to rise around 800 years after the continental temperature as an ice age draws to a close. This suggests that some other agent initiates the warming, but then CO_2 plays a role in amplifying the effect through its radiation-trapping abilities. The initial trigger would potentially be a change in the amount of summer sunshine reaching polar regions due to variations in the Earth's orbit around the sun, or a warming caused by another mechanism such as a change in the distribution of planetary heat by shifting ocean currents. The 800-year lag is around the time taken to flush out the deep ocean through natural ocean currents. Therefore it is possible that CO_2 might be stored in the ocean during ice ages, and then get released when the climate begins to warm. One piece of evidence for this is that the "carbonate compensation depth" (CCD) of the Atlantic and Pacific was shallower during the last ice age. This is the level in the oceans at which the rate of calcium carbonate supplied by shells of sea creatures equals the rate at which the material dissolves – so no calcium carbonate exists below this depth. The level depends upon temperature, pressure and chemical composition, in particular the amount of dissolved CO_2. The implication is that a shallower CCD reflects a higher concentration of CO_2 stored in the deep ocean.

Methane increase

The findings from the initial Antarctic core spurred scientists around the world to drill new, longer cores and analyze the existing ones in different

A scientist carefully prepares a new ice core for storage.

ways. In 1990, scientists at the French Laboratory for Glaciology and Geophysics near Grenoble measured the level of methane present in a core extracted at Vostok in east Antarctica. Fourteen years earlier, scientists at NASA Goddard Institute for Space Studies in the USA had discovered methane to be a significant greenhouse gas, absorbing heat in much the same way as carbon dioxide. It is produced primarily by bacteria living in swamps, so it made sense that there would be less around during ice ages when the ground was frozen, and more during interglacials. Sure enough, the concentrations preserved in air bubbles spanning the past 160,000 years showed that levels of methane were around 350 parts per billion (ppb) during past ice ages, rising to 650ppb during warmer interglacials. The core data suggested that, while changes in CO_2 took place over several thousand years, most of the methane fluctuations occurred rather quickly.

Subsequent work by John Kutzbach made sense of this by suggesting that, when Milankovitch's 22,000-year radiation cycle strengthens the Indian summer monsoon, the increased rain keeps the tropical wetlands moist, prompting a rise in methane emissions. At the end of an ice age methane increases, owing to the thaw encouraging increased bacterial activity in the same way. When an ice age begins, the gas decreases. What Kutzbach's work could not explain, however, was the scientists' other findings: that the amount of methane in the air had increased from around 700ppb in 1700 to 1,700ppb in 1990. Up to 1900 it had been rising at a rate of 1.5ppb; from 1900 this had speeded up to 17ppb per year. Although methane levels are considerably lower than CO_2, methane is 21 times more effective at absorbing radiation than carbon dioxide, so the researchers were concerned that the rise would bring disproportionately strong greenhouse warming. (New work suggests methane levels have been bucking the Milankovitch trend by rising for the past 5,000 years, which some scientists blame on human agricultural activities, such as diverting rivers to irrigate rice. However, this remains a controversial theory at this stage.)

Cores were also extracted on the other side of the world to Antarctica. In the late 1980s and early 1990s, separate European (GRIP) and USA (GISP2) projects yielded two ice cores 30 kilometres apart taken close to the summit of the Greenland ice cap. The location for the cores was chosen to be near the summit as this area would generate the longest cores. Also, as ice flows away from the centre of an ice cap it would minimize the chance of ice flow distorting the layers. These cores yielded ice dating back 110,000 years, which meant that they extended back through the current 10,000-year interglacial period (called the Holocene), through the 100,000 years of the last ice age. Unlike the Antarctic cores, which had generally low resolution owing to the very slow rate of accumulation of snow, the Greenland cores exhibited discernable annual layers representing each season's fresh snowfall. Researchers counted each year's band of snow-turned-ice and so built up an annual climate record. They then analysed the core in a number of different ways to coax out the information on past climatic conditions.

Rapid change

The cores caused a shock in the scientific world, as they showed that recent climate changes had happened a lot more quickly than previously thought. It now emerged that, as the world had moved into and out of the last ice age, abrupt climatic shifts punctuated the general cooling and warming trends. In some instances, temperature changes of a magnitude equal to half the difference between ice-age conditions and the modern-day climate had occurred in just a few years or decades. While humans had enjoyed a comparatively stable climate throughout the current interglacial, large, rapid climate oscillations had clearly been the norm during the whole of the past 100,000-year glacial period. For example, the cores revealed that modern Greenland is 20°C warmer than it once was, but that its temperature had at times risen by as much as 8°C in a matter of decades during the ice age.

Scientists gleaned more information by measuring the amounts of chemicals blown onto Greenland, such as the sodium and chlorine from sea salt, calcium from continental dust and ammonium from vegetation at lower latitudes. These studies indicated that major changes such as switches in temperature of many degrees Celsius, two-fold changes in snow accumulation, large changes in the amount of wind-blown dust and sea salt and large changes in methane concentration had taken place over decades or even less. They surmised that such dramatic changes could only be accounted for by major shifts in the circulation patterns of ocean currents and winds, and large-scale changes in global wetlands. The rapid warm and cold oscillations became known as Dansgaard/Oeschger (D/O) events, after the scientists that discovered them.

At about the same time that scientists were busily working to extract the Greenland ice cores, researcher Hartmut Heinrich at the University of Göttingen in Germany had found half a dozen unusual rock layers in sediments cored from the North Atlantic sea bed. Recorded as far apart as Canada and Bermuda, the rock bands appeared at regular intervals of around 8,000 years all the way through the last glaciation to its end 10,000 years ago. Closer scrutiny of the rocks' provenance revealed that they had come from the Hudson Bay area of northern Canada. Because some of the particles were too big to have been carried by water, Heinrich concluded that the geological debris must have been carried by armadas of icebergs which floated away from the North American ice sheet and then melted on their journey south. This is now referred to as a Heinrich event.

Gerard Bond from Lamont-Doherty Earth Observatory thought that the Heinrich events and Dansgaard/Oeschger (D/O) cycles might be linked and that, if so, they might have coincided with climate changes in other parts of the planet. He decided to try and find out by re-examining Atlantic sediment cores spanning the past 30,000 years. Not only did he find Heinrich's rock fragments every 8,000 years, but he picked up more closely spaced, less obvious, layers of rock fragments occurring roughly every 1,500 years, the same frequency as the D/O events. The rocks in these layers came from east Greenland, the Arctic Island of Svalbard and Iceland. Bond came to the conclusion that these fragments, too, had hitched a ride on southbound icebergs and had been deposited when the ice melted. More research revealed that the isotopic and chemical signatures of the events were not limited to the North Atlantic region. Signals indicating that rapid climatic switches had taken place in the past turned up in cores from Santa Barbara, California, the Cariaco Basin off Venezuela, and a location off the coast of India. In all, more than 170 locations around the planet have confirmed these cyclical changes. Whatever was happening in the North Atlantic was somehow having an impact on climate around the globe.

Subsequent work has shed new light on both the Heinrich and D/O events. Twenty-four D/O events have now been identified from the glacial period between 115,000 and 14,000 years before the present. In each case, the evidence points to a period of rapid warming, typically in a few decades, each followed by a more gradual period of cooling. The subsequent shape of the temperature profiles is saw-toothed. Each overall event lasts for between 1,000 and 3,000 years. Heinrich events occur primarily during the cold periods of some D/O cycles, are relatively brief and tend to occur at the boundaries of major climatic transitions. For example, one event marks the start of the end of the last interglacial.

Scientists now believe that the mechanisms that induce D/O and Heinrich events ultimately trigger changes in the Meridional Overturning Circulation (MOC). This is the oceanic conveyor belt that has distributed heat around the planet throughout the current interglacial period. The Gulf Stream part of the conveyor carries warm, salty water up from the tropics into the Atlantic, where an offshoot called the North Atlantic Drift continues northward to around 80° N, sinks, and then returns to the southern hemisphere at depth. This "overturning" process occurs because cold water is denser than warm water and salty water denser than fresh. The deep returning current is called the North Atlantic Deepwater Formation. Scientists believe that, during the last ice age, the sudden input of freshwater from icebergs, such as that caused by Heinrich events, diluted the saltiness of the northward-headed current, preventing the MOC from overturning and carrying cold waters back south at depth.

By studying the timings of temperature changes related to D/O and Heinrich events, scientists believe that the MOC can exist in three states. The first state occurs during warm interglacial periods, such as the present day. Large amounts of heat are carried to northern high latitudes, then deep waters form in the Nordic seas and flow back south. The second state exists during ice ages. In this scenario, deep water forms south of the shallow sill between Greenland, Iceland and Scotland, so that the amount of heat transferred to high northern latitudes is greatly reduced. The third state is a "switched off" mode, in which there is practically no deep-water formation in the Atlantic. In the last mode, the deep circulation of the Atlantic is dominated by an inflow of chilly Antarctic Bottom Water. Because no heat is being carried from the tropics up to northerly latitudes, northern Europe rapidly cools.

Model simulations suggest that the second mode is the only state that is stable during glacial conditions. D/O events can therefore be explained as incursions of warm Atlantic waters into the Nordic seas that shift the deep-water formation north when increases in the Earth's solar output cause the ocean conveyor to switch into the warm mode. Because the MOC is not stable in this new state under glacial conditions, it gradually reverts to the cold stadial state over a thousand or so years. During the cold stages of some D/O events, the switch operates the other way and stops the MOC circulation altogether. The cause in this case is the occurrence of a Heinrich event. The implication is that something suddenly causes an armada of icebergs to calve away from the polar ice sheet. The cause is not yet understood, but it could simply be that ice builds up until it reaches some kind of mechanical breaking point, at which time the sheet begins to disintegrate. Whatever the cause, the rapid input of freshwater from the melting icebergs reduces the density of the ocean and stops the water from sinking and flowing back south. This switched-off state is also unstable during glacials, so the circulation slowly reverts to the second mode over a period of time.

While D/O and Heinrich events show that the MOC can switch between states, they are essentially glacial processes. However, scientists have also identified two rapid cooling phases in the palaeoclimatic record, akin to Heinrich events, in which the oceanic conveyor switched off while the world was undergoing a period of warming. The first is the Younger Dryas event, which took place between 12,900 and 11,500 years ago and is named after the hardy Arctic flower *Dryas octopetala*, whose favoured habitat is frigid

tundra. At the start of the Younger Dryas, conditions had already warmed to near-interglacial temperature when the climate in the North Atlantic region switched back into glacial conditions in a matter of decades. This cold spell prevailed for more than a thousand years. Scientists believe that the waters of a huge meltwater pool on the North American continent, Lake Agassiz, broke through an ice barrier and flooded into the North Atlantic via the St Lawrence river. The second period of abrupt cooling took place some 8,200 years ago and lasted for around 150 years. Its cause was again likely to be a rapid discharge of fresh meltwater into the North Atlantic, this time via Hudson Bay. On both occasions the MOC stopped, plunging the northern hemisphere into icy conditions that had a knock-on effect on climate across the whole world.

It is possible that melting Arctic ice could reduce the salinity and density of water in the North Atlantic sufficiently to halt the Meridional Overturning Circulation.

Pressure points

So what does all this have to do with today's warming world? In the film *The Day after Tomorrow* (Roland Emmerich, 2004), much of the northern hemisphere is plunged almost instantaneously into an ice age when a rapid influx of freshwater from melting Arctic ice prompts the northern part of the MOC, the North Atlantic Drift, to break down. Although the speed of change in the film is greatly exaggerated, it is based on the science outlined above and may not be as far-fetched as it seems. There is a possibility that unusually large releases of freshwater from melting Arctic ice and increased rainfall could dilute the North Atlantic Drift again,

The world's tipping points

In 2004, Professor John Schellnhuber, then research director of the Tyndall Centre for Climate Change in Norwich, UK, identified several planetary tipping points, where global warming could trigger sudden catastrophic changes.

1. GREENLAND ICE SHEET

Some 2.6 million cubic kilometres of water are locked up in the Greenland ice sheet. Scientists have predicted that if the temperature rises by 3°C a slow thaw will set in, releasing water into the North Atlantic. A rise of 8°C could melt the entire sheet, prompting a seven-metre rise in sea level that would inundate coastal communities around the globe. A study conducted by the University of Texas at Austin, USA, in 2006 found the ice sheet was melting three times faster than it had been five years before. Researchers concluded that it was losing 200 cubic kilometres annually, and contributing 0.5 millimetres to sea-level rise each year.

2. AMAZON RAINFOREST

One of the most biodiverse regions on Earth, the Amazon rainforest is also a vast store of carbon. Models suggest global warming will cause a drop in Amazon rainfall, leading to frequent droughts. The result would be the slow death of the forest, the extinction of thousands of species and a massive release of carbon into the atmosphere. One study has already suggested that, after two years of drought and decades of logging, the forest is already approaching its tipping point, after which it will die and slowly turn into a vast desert.

5. EL NIÑO IN THE PACIFIC OCEAN

The natural phenomena of El Niño warms the waters of the Pacific, shifting weather patterns. Models have predicted that El Niño events could become more severe and frequent, bringing drought to Southeast Asia and northwest Australia and flooding parts of South America.

3. WEST ANTARCTIC ICE SHEET

Because Antarctica's ice is so thick, scientists had not thought it would be vulnerable to rapid melting in the wake of global warming. However, in 2002, the Larsen-B ice shelf broke off the eastern side of the Antarctic Peninsula and deposited 3,250 square kilometres of ice into the sea in fragmented icebergs. British Antarctic Survey has since reported that it had found evidence linking the collapse of the shelf to human activity.

4. NORTH ATLANTIC DRIFT

The North Atlantic Drift branches off from the Gulf Stream, which carries warm water from the tropics to northerly latitudes. In the North Atlantic, the surface waters cool, sink to the ocean floor and head back towards the equator. This is the Meridional Overturning Circulation (MOC). The current delivers the equivalent of 100,000 large power stations' worth of free heating, making northern European countries much warmer than their Canadian counterparts at equal latitudes. Global warming could change that by increasing the freshwater entering the North Sea from melting ice and greater rainfall. The IPCC's A1B scenario forecasts the current could slow-down by as much as 50 per cent in the twenty-first century.

6. THE ANTARCTIC CIRCUMPOLAR CURRENT

The Antarctic Circumpolar Current circulates 140 million cubic metres of water per second. There are also currents that act vertically, upwelling in places to bring nutrients to the surface and downwelling where the water is cold, salty and dense. Climate change is expected to bring more rainfall to the southern polar region, which could slow the rise of nutrients, cutting food supplies of marine species.

NORTH AMERICA

SOUTH AMERICA

ATLANTIC OCEAN

7. SALINITY VALVES, GIBRALTAR

In some places the local geography causes the water between adjacent seas to be squeezed into discrete water bodies. If one body of water is saltier than the one next to it, a flux of salt nutrients and oxygen can form across the gap, producing a salinity valve. This is the case at Gibraltar. The gradients across the valve create unique ecosystems that become well-adapted to the conditions. Global warming could disrupt the water flows, bringing conditions intolerable to the delicate flora and fauna.

9. METHANE CLATHRATES, SIBERIA

The Siberian permafrost and ocean-floor sediments contain vast deposits of gas-filled ice, called clathrates. At Siberia's chilly temperatures and under the pressure of icy oceans clathrates are stable, but if the planet warms up it is possible that melting ice will relinquish its methane. Scientists say there may be as much as 10–11 trillion tonnes of carbon held in clathrates, 20 times the amount held in natural gas reserves. Because methane is a strong greenhouse gas, such a release could increase global warming by up to 25 per cent. Researchers have since discovered that a million square kilometres of ground has begun melting for the first time since forming 11,000 years ago, at the close of the last ice age.

EUROPE

ASIA

9

8

11

10

AFRICA

8. TIBETAN PLATEAU, CHINA

The vast Tibetan Plateau, a quarter of China's landmass, is permanently covered by snow. Its white blanket helps reflect the sun's rays back into space, so keeping the region cool. If the ice in this region melts, the darker greys and browns of the exposed landmass will absorb much more heat, accelerating melting. In 2006, the Chinese Academy of Sciences reported that the glaciers of the Tibetan Plateau were melting so fast they were halving in volume every 10 years. The plateau has 46,298 glaciers. Each year, enough water is melting to fill the entire Yellow River.

10. INDIA'S MONSOON

India's weather heats up throughout March and April, until temperatures peak in May. The parched land generates a sharp temperature gradient between land and sea, prompting the winds to switch rapidly from seaward to landward. These winds release their burden of moisture of monsoon deluges as they cross the land. Global warming appears to be strengthening the monsoon. There has been a significant increase in the magnitude and frequency of freak rain events.

12. OZONE HOLE, ABOVE ANTARCTICA

Although scientists predict the hole in the ozone layer will close up within half a century, climate change could undo the good work of the Montreal Protocol. Signed in 1987, the agreement banned the use of CFCs that had created the hole, leading to its successful shrinkage. Global warming could cause cooling in the stratosphere where ozone forms. If this happens, it will disrupt the chemical process that prevents ozone from breaking down, so, as the world warms up, ozone will be lost. It is possible the hole would open over Europe, increasing the risk of skin cancer and blindness.

11. SAHARA DESERT, AFRICA

Parts of the Sahara could become greener as rainfall soaks its southern margins. While beneficial for communities living in newly fertile areas this may have a negative impact on other parts of the planet. Currently, Saharan dust is swept up into clouds and carried across the Atlantic where 200 million tonnes fall into the ocean. The dust provides nutrients that generate blooms of plankton, on which the marine food chain relies. A greener Sahara would produce less dust, cutting the foundations of the ocean food supply. Also, because plankton absorbs CO_2, less dust would make the ocean less effective at extracting the greenhouse gas from the air, causing more warming.

12

reducing its density and slowing it or stopping it from sinking. If the sinking stopped, this would effectively switch off the global current. Tropical waters would no longer flow past the coast of Europe, and the climate would cool rapidly.

In 2006, the findings of a study that compared past temperatures in Greenland and Antarctica were published in the scientific journal *Nature*. Scientists analysed the ratio of ^{18}O and ^{16}O isotopes from one core taken from Queen Maud Land as part of the European Project for Ice Coring in Antarctica (EPICA) and one extracted on the opposite side of the world by researchers from the North Greenland Ice Core Project (NGRIP). The results suggested that, during the last ice age, when it was chilly in Greenland, it got warmer in Antarctica. This implied that when a Heinrich event injected sufficient freshwater into the North Atlantic to switch the MOC off, the lack of cold returning water prompted Antarctica to gradually warm, and when the ocean slowly reverted to its stable state by switching the conveyor back on, Antarctica started to get colder. This was strong evidence supporting the theories that the ocean could exist in a different state from its present one, and that it had done so in the past.

A survey conducted in 2002 found that rivers flowing into the Arctic Ocean were discharging seven per cent more water than they did in the 1930s. Three years later, a study of ocean circulation concluded that the warm water currents carrying water north from the Gulf Stream had reduced by 30 per cent. At this stage it is not certain whether this reflects a long-term change, or natural variability. However, in 2006, researchers reported that the Greenland ice sheet was losing a cubic kilometre of water every 40 hours from icebergs crashing into the Atlantic. This is equal to the amount of water Los Angeles consumes in a year. The total annual loss, after allowing for snowfall, was 220 cubic kilometres, more than twice the amount lost in the mid-1990s. Computers predict that it is very likely that the MOC will slow down during the twenty-first century, with best estimate being for a 25 per cent slow-down by 2100. Whether the warming temperature will override the potential cooling effect is as yet uncertain. Much of the ongoing work into climate change is concerned with understanding what causes rapid switches in climate and what set of circumstances might cause the MOC to slow or cease in future.

A primary lesson learned from detailed studies of the MOC is that the global natural processes that control our climate are capable of flipping from one state to another within the span of a human lifetime. Scientists have now identified twelve regions of the globe that may act as "tipping points" to switch the world into new climatic states if pushed too far by human activities overloading the atmosphere with heat-trapping green-house gases. In addition to the North Atlantic Drift, they include the Greenland ice sheet, the Amazon rainforest, the Tibetan plateau, the Sahara Desert and the ozone hole. The idea was put forward at EuroScience Open Forum 2004 in Stockholm, by Professor John Schellnhuber, currently Director of Potsdam Institute for Climate Impact

Research (see map, page 28). Since then, the Amazon has suffered a pro-longed drought from which it may not recover, the glaciers of the Tibetan Plateau have begun melting at a rate of seven per cent per year, and the apparent northward shift of the Sahara has repeatedly brought drought to the Mediterranean. Meanwhile melting of the Greenland and West Antarctic Ice Sheets has accelerated.

We are responsible

The Intergovernmental Panel on Climate Change (IPCC) has the job of gathering and disseminating information on climatic changes taking place around the world. Set up in 1988 by the World Meteorological Organization (WMO) and the United Nations Environment Programme (UNEP), its global amalgamation of experts comprises three working groups. These address: the scientific aspects of the climate system and climate change; the impacts of and adaptations to climate change; and the options for the mitigation of climate change. The overall aim is to "provide an assessment of the understanding of all aspects of climate change including how human activities can cause such changes and can be impacted by them". So far the panel has released four reports, in 1990 (First Assessment Report), 1996 (Second Assessment Report [SAR]), 2001 (Third Assessment Report [TAR]) and 2007 (Fourth Assessment Report [FAR]). In the early days no one really knew whether human activities were truly the cause of the observed warming trend and elevated CO_2 levels. But as time went by it became more and more obvious that humans were the culprits and were beginning to take control of the Earth's climate.

This increasing understanding of the scale of human influence is reflected in the changing language used in the reports. Whereas the 1996 publication reported: "the balance of evidence suggests a discernible human influence on global climate", the 2001 publication said that "there is new and stronger evidence that most of the warming observed over the last 50 years is attributable to human activities". The 2007 report leaves no room for doubt about the cause of global warming: "Most of the observed increase in globally averaged temperatures since the mid-twentieth century is very likely [more than 90 per cent certain] due to the observed increase in anthropogenic [caused by humans] greenhouse gas concentrations". The FAR confirms that the increase in the global CO_2 concentration from a pre-industrial level of 280ppm to 370ppm in 2005 has pushed up the temperature by 0.76°C.

Although our knowledge of climate change science has improved greatly in the past decades, there are still many uncertainties about what changes a warmer world might unleash. So we could be in for some climatic surprises in the coming decades. In 2001, the TAR stated that the thermal expansion of seawater warmed by the enhanced greenhouse effect was likely to contribute the largest component to sea-level rise over the next 100 years. Changes in sea level take place on timescales from hours (tides) to millions of years (changes to ocean basins caused by geological processes). On a scale of decades to centuries, climate has a major influence. In the past 3,000 to 5,000 years, fluctuations in global sea level on timescales of 100 to 1,000 years are unlikely to have breached 50 centimetres. And during the twentieth century the rate of global mean sea-level rise was very small, between one and two millimetres per year. But at the last glacial maximum, 20,000 years ago, sea level was 120 metres lower than it is today. The TAR predicted that the melting of mountain glaciers and ice caps, which make up only a few per cent of the world's land-ice area, would add to the amount of water in the ocean. But it considered that the large ice sheets of the Arctic and Antarctic, which contain a considerable proportion of the world's freshwater, would make only a small net contribution to sea-level change in the coming decades.

Only a year after the report was published, however, Antarctica's 3,250-square-kilometre Larsen B Ice Shelf collapsed. British Antarctic Survey later reported that it had the first direct evidence linking human activity to the collapse of the shelf. It said that global warming and the ozone hole have changed Antarctic weather patterns such that strengthened westerly winds now force warm air eastward over the natural barrier created by the Antarctic Peninsula's two-kilometre-high mountain chain. When this happens in summer, temperatures in the north-east Peninsula warm by around 5°C, creating the conditions that prompted meltwater to drain into crevasses on the Larsen Ice Shelf, a key process that led to its break-up. In February 2006, the UK Government released a report saying that the temperature of Greenland had climbed to a point such that its ice sheet's tipping point was not far off. In an interview with *New Scientist*, the director of NASA Goddard Institute for Space Studies, Jim Hansen, said the collapse of the sheet could be "explosively rapid", generating sea level rises of a couple of metres within a century and several more during the next century.

The 2007 FAR adjusted the IPCC's projections for sea-level rise down to between 18 and 59 centimetres, but these new forecasts did not take into account possible changes in rates of melting to the Greenland and Antarctic ice sheets. Melting of these sheets remains a possibility, but computer models are not currently capable of modelling accurately the possible outcomes, hence the exclusion. More than 300 million people worldwide live within one metre of average sea level, and of the world's 15 largest cities 13 are on coastal plains. Many heavily populated areas, for example Bangladesh and the Netherlands, are near or below sea level. A sudden rise in sea level would have disastrous consequences for coastal dwellers.

One extreme to another

One prediction that has long been made by scientists is that climate change will increase the frequency of extreme events. And this certainly appears to be happening. In recent decades, locations around the globe have reported an increasing number of unusual weather conditions. They range from the European heat wave of 2003 that killed more than 30,000 people, to the floods that submerged two-thirds of Bangladesh the following summer, and the powerful Hurricane Katrina that flattened New Orleans in 2005. Munich Re, one of the world's largest reinsurance companies, estimates that the world suffered more than US$210 billion in economic losses as a result of weather-related natural disasters in 2005, making it the costliest year on record. The cumulative number of people affected by disasters rose to two billion in the 1990s, up from 740 million in the 1970s. Virtually all of these millions were concentrated in poorer regions.

The bottom line

In late 2006, a report by economist Sir Nicholas Stern suggested that global warming could shrink the global economy by as much as 20 per cent but that taking action now would cost just one per cent of

Sir Nicholas Stern's 2006 report suggested climate change could shrink the global economy by as much as 20 per cent.

global gross domestic product. The 700-page report said that, without action, floods from rising sea levels could displace up to 200 million people by the middle of the century; melting glaciers could cause water shortages for one-sixth of the world's population, droughts may create tens or even hundreds of millions of "climate refugees" and up to 40 per cent of species of wildlife could become extinct with 2°C of warming. "Whilst there is much more we need to understand – both in science and economics – we know enough now to be clear about the magnitude of the risks, the timescale for action and how to act effectively", Sir Nicholas said. His words are a clear warning that unless we all make efforts to reduce greenhouse gas levels and reduce the impact of the enhanced greenhouse effect, we will gain first-hand experience of the planet's ability to switch irreversibly into an entirely new climatic regime.

ARCTIC OCEAN

The human cost of climate change

An increasing number of extreme weather events, which most scientists agree are consistent with climate change, are taking their toll on communities around the world.

3. CANARY ISLANDS

2005: Tropical Storm Delta was the first-ever tropical storm to strike these islands. Meanwhile, Hurricane Vince was the first-ever hurricane to approach continental Europe, making landfall in Spain.

NORTH
AMERICA
1

5

1. NORTH AMERICA

2005: Hurricane Katrina struck Louisiana and Mississippi, causing one of the most destructive disasters in US history. Hurricane Rita also caused severe damage in Louisiana and Texas.

3

ATLANTIC

OCEAN

PACIFIC OCEAN

SOUTH
AMERICA
2

2. SOUTH AMERICA

2005: Hurricane Stan took 2,000 lives, while Hurricane Wilma devastated Yucatan, Mexico, before striking Florida. The year smashed all records for the Atlantic hurricane season. By 31 December, there had been 27 tropical storms (six more than the previous record of 1933), and 14 hurricanes, breaking the 1969 record of 12 hurricanes.

4. AFRICA

2005: Algeria experienced its heaviest snowfall for half a century. Persistent drought, alternated by floods, affected Kenya and Ethiopia. Food shortages linked to climate change were reported to be forcing fathers to push their daughters to marry well before their teens in order to acquire valuable livestock as a dowry.

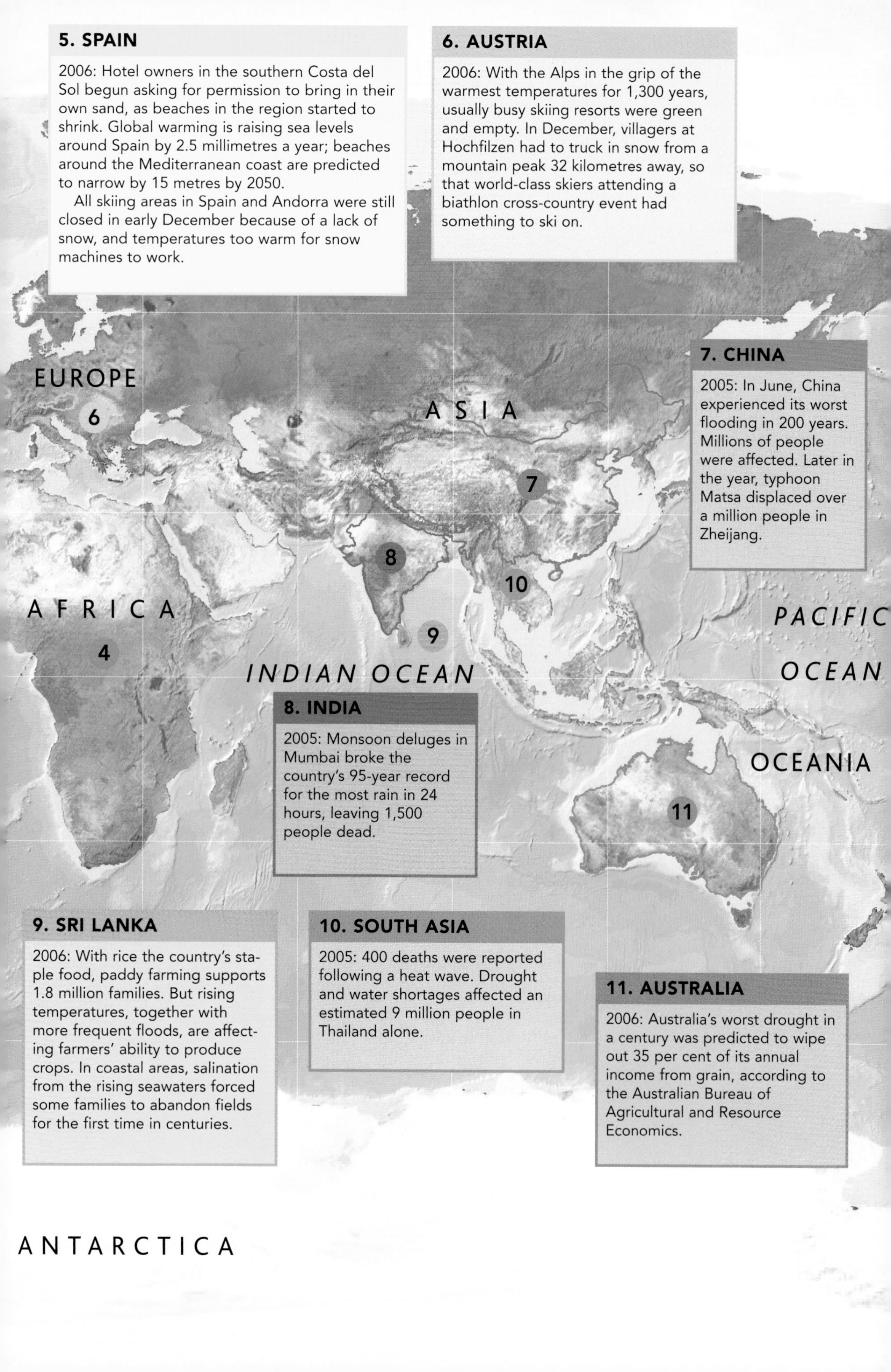

5. SPAIN

2006: Hotel owners in the southern Costa del Sol begun asking for permission to bring in their own sand, as beaches in the region started to shrink. Global warming is raising sea levels around Spain by 2.5 millimetres a year; beaches around the Mediterranean coast are predicted to narrow by 15 metres by 2050.

All skiing areas in Spain and Andorra were still closed in early December because of a lack of snow, and temperatures too warm for snow machines to work.

6. AUSTRIA

2006: With the Alps in the grip of the warmest temperatures for 1,300 years, usually busy skiing resorts were green and empty. In December, villagers at Hochfilzen had to truck in snow from a mountain peak 32 kilometres away, so that world-class skiers attending a biathlon cross-country event had something to ski on.

7. CHINA

2005: In June, China experienced its worst flooding in 200 years. Millions of people were affected. Later in the year, typhoon Matsa displaced over a million people in Zheijang.

EUROPE

6

ASIA

7

AFRICA

4

8

10

INDIAN OCEAN

9

PACIFIC

OCEAN

OCEANIA

11

8. INDIA

2005: Monsoon deluges in Mumbai broke the country's 95-year record for the most rain in 24 hours, leaving 1,500 people dead.

9. SRI LANKA

2006: With rice the country's staple food, paddy farming supports 1.8 million families. But rising temperatures, together with more frequent floods, are affecting farmers' ability to produce crops. In coastal areas, salination from the rising seawaters forced some families to abandon fields for the first time in centuries.

10. SOUTH ASIA

2005: 400 deaths were reported following a heat wave. Drought and water shortages affected an estimated 9 million people in Thailand alone.

11. AUSTRALIA

2006: Australia's worst drought in a century was predicted to wipe out 35 per cent of its annual income from grain, according to the Australian Bureau of Agricultural and Resource Economics.

ANTARCTICA

THE ATMOSPHERE

The Earth is surrounded by a blanket of air called the atmosphere. Stretching for some 560 kilometres from the surface of the planet, it is primarily made up of nitrogen (N_2, 78 per cent), oxygen (O_2, 21 per cent), and Argon (Ar, 1 per cent). There are other influential components present in small amounts. These include water (H_2O, 0–7 per cent), ozone (O_3, 0–0.01 per cent) and carbon dioxide (CO_2, 0.01–0.1 per cent). Viewed from space, the atmosphere appears as a fragile, enclosing membrane.

❶ The troposphere is thicker at the equator and thinner at the poles. It contains almost all our weather. The atmosphere becomes less dense with distance from the Earth's surface. At the top of Mount Everest, almost nine kilometres up, the atmospheric pressure is 33 per cent of that at sea level. This means that, although the percentage of oxygen in the air is still 21 per cent, there is 66 per cent less oxygen than at sea level. Because of this, high-altitude climbers have to acclimatize themselves to the conditions or risk becoming ill. Kerosene will not burn at this altitude and helicopters cannot fly.

❷ The troposphere and stratosphere together contain 99 per cent of our "air". The stratosphere is drier and less dense than the troposphere. Importantly, it contains the ozone layer. Ozone is a molecule containing three oxygen atoms (O_3). Normal oxygen, that we breathe, has two oxygen atoms (O_2). Ozone is much less common than normal oxygen. Out of each 10 million air molecules, about two million are normal oxygen, but only three are ozone. Even this small amount plays a valuable role in shielding the Earth from harmful radiation. Most importantly, it absorbs the portion of ultraviolet light called UVB. This type of energy has many harmful effects, including causing skin cancers and cataracts, and destroying crops.

❸ In the mesosphere molecules are in an excited state, as they readily absorb energy from the sun. The gases, including oxygen, get thinner and thinner with height, so the temperature drops with altitude to –100°C. This is colder than the lowest temperatures on Earth and it freezes water vapour into the "noctilucent" clouds that we sometimes see at sunset. Although greatly reduced, the gases in the mesosphere are thick enough to slow down meteorites hurtling into the atmosphere, where they burn up. We recognize their fiery trails in the night sky as shooting stars.

❹ The gases of the thermosphere are increasingly thinner than in the mesosphere. Only the higher energy ultraviolet and x-radiation from the sun are absorbed. This absorption causes the temperature to increase with height; near the top it can be as much as 2,000°C. However, despite the high temperature, this layer of the atmosphere would still feel very cold to our skin because of the extremely thin air. The total amount of energy emitted from the very few molecules present in this layer is not sufficient to heat our skin. The absorption of energy from the sun also gives rise to the ionosphere. Solar radiation, acting on the different compositions of the atmosphere with height, generates layers of electrically charged, or ionized, gas particles within the thermosphere. The ionosphere allows us to bounce radio signals from one point on Earth and receive them at another. Above the thermosphere lie the thermopause and exosphere. In this outermost layer, atoms and molecules escape into space and satellites orbit the Earth.

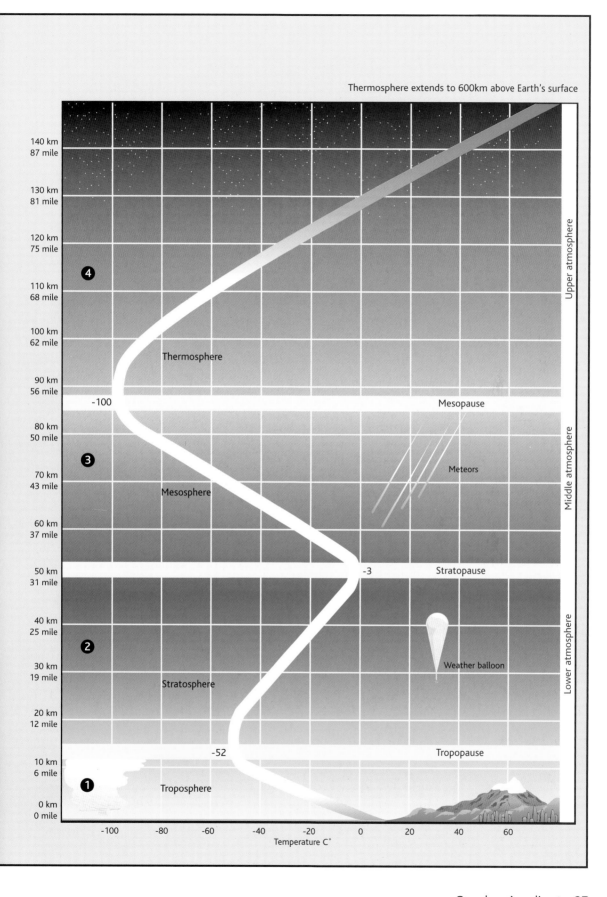

Thermosphere extends to 600km above Earth's surface

140 km
87 mile

130 km
81 mile

120 km
75 mile

4

110 km
68 mile

100 km
62 mile

90 km
56 mile

Thermosphere

-100   Mesopause

80 km
50 mile

3

70 km
43 mile

Meteors

Mesosphere

60 km
37 mile

50 km
31 mile

-3   Stratopause

40 km
25 mile

2

Weather balloon

30 km
19 mile

Stratosphere

20 km
12 mile

10 km
6 mile

-52   Tropopause

1

Troposphere

0 km
0 mile

-100   -80   -60   -40   -20   0   20   40   60

Temperature C°

Upper atmosphere

Middle atmosphere

Lower atmosphere

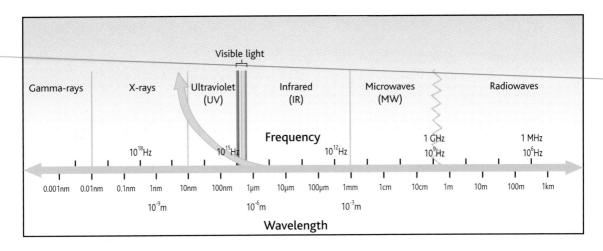

This spectrum shows the complete range of wavelengths of electromagnetic radiation. The sun emits radiation within the ultraviolet, visible light and infrared regions.

The greenhouse effect

The greenhouse effect is well understood. The sun radiates large quantities of energy across a wide range of wavelengths. The shorter the wavelength, the higher the energy emitted. Between seven and eight per cent of the energy comes from wavelengths that are shorter than those of visible light, such as ultraviolet light. A much larger percentage of energy, around 43 per cent, comes from visible wavelengths between 400 and 700 nanometres (1,000 nanometres = 1 micrometre) as sunlight. The remaining 49–50 per cent is radiated at wavelengths longer than visible light. These lie in the near infrared range from 700 to 1,000 nanometres; the thermal infrared (heat), between 5 and 20 micrometres; and the far infrared regions.

Most of the energy emitted at the shortest wavelengths is absorbed by ozone in the stratosphere. Various components of the Earth's atmosphere also absorb the infrared solar radiation before it penetrates to the surface, but the atmosphere is quite transparent to visible light so this passes through towards the Earth's surface. Around a third of the incoming sunlight is reflected back to space by clouds, particles in the air and the Earth's surfaces. Pale surfaces, such as ice fields, reflect more sunlight back than darker ones. Another 20 per cent is absorbed by clouds and water vapour before reaching the ground. The remaining half of the incoming sunlight is absorbed by the Earth's surfaces such as forests, deserts, city roads and buildings, and the oceans.

The energy absorbed by the Earth is later emitted as thermal infrared energy. Greenhouse gases such as CO_2 and methane absorb the thermal infrared energy emitted by the Earth's surface, so the atmosphere warms. Before the industrial revolution, the natural greenhouse effect kept the Earth 33°C warmer than it would have been with no atmosphere. By comparison, the moon, which has no atmosphere, has an average surface temperature of –18°C. The additional injection of greenhouse gases from human activities since the start of the industrial age has pushed up the Earth's temperature by a further 0.76°C.

Warming agents: greenhouse gases

Carbon dioxide

CO_2 has been in the atmosphere for over 4 billion of Earth's 4.5 billion years. In Earth's early history, atmospheric concentrations of CO_2 peaked at around 80 per cent, but most of the gas was removed from the atmosphere as early organisms evolved the process of photosynthesis. When the organisms died, the carbon they had removed from the atmosphere became locked up as carbonate minerals in rocks (the chalky white cliffs of Dover in the UK are one example), plus geological coal, oil and gas formations.

A natural cycle constantly processes carbon in various forms through atmosphere, oceans, vegetation, soils and rocks. This happens through a system of fluxes of CO_2 between land (vegetation and soils), ocean (water and ecosystems) and the atmosphere. CO_2 is removed from the atmosphere primarily by photosynthesis, the process by which plants build cells of organic carbon using CO_2 from the atmosphere and energy from the sun. CO_2 also dissolves in the surface waters of the oceans. It is returned to the atmosphere by respiration, the process used by living things to turn

How the greenhouse effect warms the Earth.

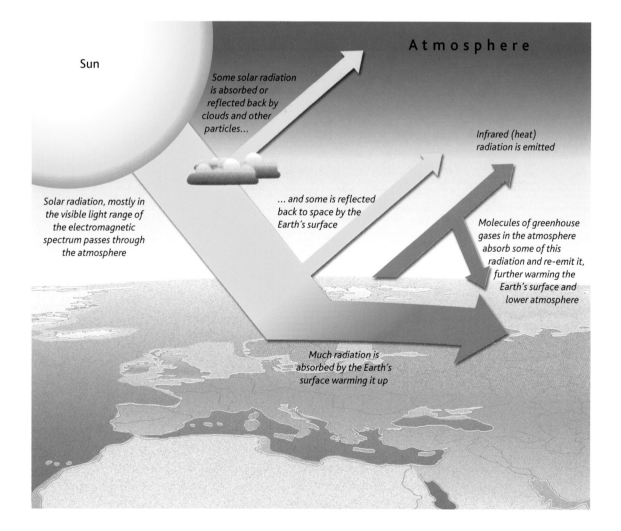

Sun

Some solar radiation is absorbed or reflected back by clouds and other particles...

Atmosphere

Infrared (heat) radiation is emitted

Solar radiation, mostly in the visible light range of the electromagnetic spectrum passes through the atmosphere

... and some is reflected back to space by the Earth's surface

Molecules of greenhouse gases in the atmosphere absorb some of this radiation and re-emit it, further warming the Earth's surface and lower atmosphere

Much radiation is absorbed by the Earth's surface warming it up

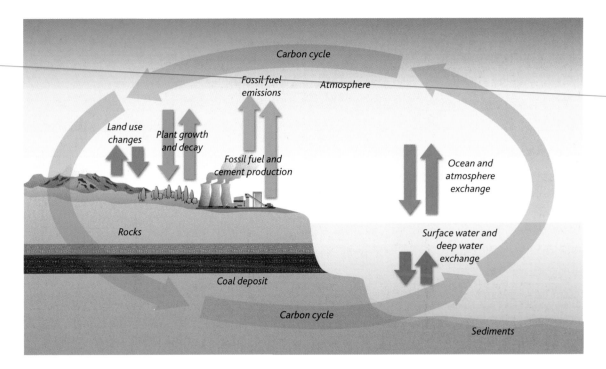

Carbon cycle
Fossil fuel emissions
Atmosphere
Land use changes
Plant growth and decay
Fossil fuel and cement production
Ocean and atmosphere exchange
Rocks
Surface water and deep water exchange
Coal deposit
Carbon cycle
Sediments

Carbon is constantly cycled through the land, ocean and atmosphere.

organic carbon plus oxygen back into CO_2. Other sources include volcanic eruptions and decaying plants.

The natural carbon cycle is balanced, so the amount taken out of the atmosphere by plants equals that returned to the air by respiration and decay. Before industrialization the concentration of CO_2 in the atmosphere was stable for thousands of years. But since industrialization, human activities have been adding to the natural equilibrium. Burning the fossil sources of carbon stored millions of years ago to provide fuel and energy, making cement from stores of chalky rocks and clearing forests all release CO_2 to the atmosphere. The result is that emissions of CO_2 from human activities have become involved in the natural cycle. Since the late 1700s, atmospheric CO_2 has risen by 35.4 per cent. Our activities are presently adding 7.2 gigatonnes of carbon (26.4 gigatonnes CO_2) per year through the burning of fossil fuels and around 1.6 gigatonnes of carbon (5.9 gigatonnes CO_2) from land-use changes such as deforestation.

Only 58 per cent of the CO_2 added by human activities remains in the atmosphere. The other 42 per cent is absorbed by the oceans, vegetation and soils. The ocean is thought to take up about 20–35 per cent, leaving 5–20 per cent to be absorbed by vegetation and soils. The atmospheric concentration of CO_2 has risen from 280ppm in 1750 to 379ppm in 2005 (see diagram right). This concentration has not been exceeded during the past 650,000 years and possibly not during the past 20 million years. It is continuing to rise at around 2ppm per year. Once added to the atmosphere, CO_2 lingers there for about a hundred years. It is the single most important infrared-absorbing anthropogenic gas in the atmosphere, accounting for 53 per cent of the total radiative forcing (see page 46) of the Earth by long-lived greenhouse gases.

Methane

Methane (CH_4) has the second-largest influence on the greenhouse effect after carbon dioxide. Even though it occurs in lower concentrations than CO_2, it produces 21 times as much warming and accounts for around 17 per cent of the "enhanced greenhouse effect". One source of natural methane is bacteria breaking down organic matter in low-oxygen environments such as swamps, and in the guts of termites and other animals. Although small, termites contribute around 20 million tonnes to the global methane concentration each year. Other natural sources are oceanic processes, and seepage from natural-gas deposits. Scientists recently discovered that plants, as well as animals, release methane and may account for between 10 and 30 per cent of global methane production. Together, these natural sources represent 40 per cent of methane emitted to the atmosphere.

The other 60 per cent of global methane comes from human-made sources. Rice paddies release large amounts, as do ruminant livestock. A single cow can belch and fart between 200 and 600 litres of methane a day. The gas also flows into the atmosphere when coal, oil and gas are extracted and transported; in the 1990s, some six per cent of methane piped across Russia was lost as a result of leaks. Other human sources include landfill sites and waste such as slurry heaps. When methane comes into contact with oxygen, it breaks down to form CO_2. It is removed from the atmosphere by chemical reactions with hydroxyl radicals (OH), which

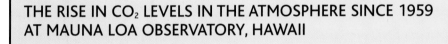

THE RISE IN CO_2 LEVELS IN THE ATMOSPHERE SINCE 1959 AT MAUNA LOA OBSERVATORY, HAWAII

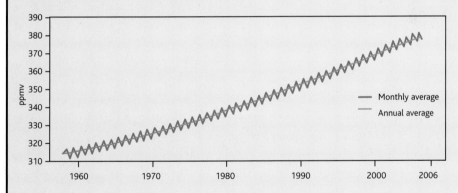

The graph displays in a zig-zag pattern because levels of CO_2 go up and down once a year. The vast majority of the world's vegetation is in the northern hemisphere, so when the northern hemisphere is tilted towards the sun in spring and summer, the leaves come out and absorb CO_2 as they photosynthesise. At this time, the level of atmospheric CO_2 decreases around the world. When the northern hemisphere is tilted away from the sun in autumn and winter, the leaves drop off the plants and release CO_2. At this point the levels rise back up again.

form when ozone is broken down by sunlight. Methane's chemistry also indirectly affects climate by influencing levels of water in the stratosphere. It stays in the atmosphere for a decade or so.

The concentration of methane in the atmosphere before the industrial revolution was around 715ppb, but that climbed to 1,732ppb in the 1990s and is now around 1,780ppb. This figure far exceeds the natural range of 320–790ppb of the last 650,000 years. After rising rapidly for a the last few decades, the amount of methane entering the atmosphere slowed in the 1990s and levelled off around the year 2000. A team of international scientists reported in 2006 that the stabilization had been prompted by two processes. First, climate change had caused wetlands to slowly dry out during the 1990s, reducing the amount of methane being released from this natural source. Second, significantly higher releases of human-made methane came from booming Asian countries, particularly China, balancing out the reduced methane coming from wetlands. The fear is that, if the drying trend is reversed, then methane levels will rise again.

Ozone

Ozone (O_3) is an important natural component of the atmosphere. It is at its greatest concentration in the stratosphere, where it is constantly being made and destroyed. It is created by interactions between oxygen and high-energy solar rays, and destroyed by the action of low-energy rays. In the stratosphere, ozone plays an important part in protecting life on Earth from harmful ultraviolet radiation from the sun. Ozone passes naturally into the troposphere and it is also created at this lower level of the atmosphere by photochemical reactions. Higher levels of air pollution in recent decades have generated an increasing amount of tropospheric ozone. Rather than being emitted by fossil fuels, it is formed by atmospheric chemical reactions between human-made emissions of hydrocarbons, nitrogen oxides (NOx), carbon monoxide and methane. At this level of the atmosphere it acts as a direct greenhouse gas as well as indirectly controlling the length of other greenhouse gas lifetimes. The breakdown of tropospheric ozone by sunlight leads to the production of hydroxyl (OH) radicals which help to "mop up" other greenhouse gases, such as methane (CH_4). It is also absorbed by plants.

Ozone remains in the atmosphere for a relatively short time, ranging from weeks to months, and changes in its concentration vary spatially. These characteristics make it much harder for scientists to estimate ozone's role in warming the planet than that of more long-lived, well-mixed greenhouse gases. Losses of stratospheric ozone, through depletion of the natural ozone layer by human-made halocarbons, have had some cooling effect in recent decades, while increased tropospheric ozone has a warming effect. The latter is offset to some extent by ozone's role in reducing methane. The current estimate for global tropospheric ozone is around 34ppb. As a direct greenhouse gas it is believed

THE OZONE HOLE

Ozone is constantly being made and destroyed in the stratosphere, keeping concentrations in equilibrium. There is some natural variation linked to levels of solar output, the seasons and latitude, but these processes are well understood and predictable. However, in 1985, scientists working in Antarctica realized that at certain times of year ozone was becoming depleted over the southern hemisphere. The culprits were found primarily to be chlorofluorocarbons (CFCs), a common industrial product used in fridges and freezers, air conditioners, aerosols and solvents. These human-made chemicals, composed of carbon, chlorine and fluorine, were found to break down ozone in the presence of ultraviolet light. Accordingly, the production of CFCs was restricted by the Montreal Protocol, which became effective in 1987. In October 2006, measurements made by the European Space Agency's Envisat satellite revealed the greatest ever ozone loss of 40 million tonnes over the South Pole. Despite the loss, scientists predict that the hole will heal fully over the next 60 or so years.

to have caused 13 per cent of all greenhouse gas warming seen since the industrial revolution. This makes it the third most important greenhouse gas after CO_2 and methane.

Nitrous oxide

All life needs nitrogen in order to live. Nitrogen is processed in a natural cycle through the atmosphere and soils, where it undergoes many complicated chemical and biological changes (including conversion to nitrous oxide, N_2O), before being returned to the air and soils. Excess nitrogen from human activities, such as the use of nitrogen-based fertilizers, emissions from human and animal waste and vehicle exhausts, now swamps the natural cycle; more than one-third of all nitrous oxide emissions comes from such activities. The result of this is that the atmospheric concentration of nitrous oxide has increased by 46ppb (17 per cent) since 1750, with present levels not exceeded for 1,000 years. Although nitrous oxide is one thousand times less abundant than CO_2 in the atmosphere, it is around 270 times more effective at trapping heat. Scientists estimate that it is responsible for around 12 per cent of the total warming from all the long-lived and globally mixed greenhouse gases. In the stratosphere, nitrous oxide destroys the ozone that provides a protective shield against the sun's harmful rays.

Halocarbons and other synthetic compounds

Halocarbons are compounds containing carbon, halogens such as chlorine, bromine and fluorine, and sometimes hydrogen. This group of gases primarily comprise CFCs (chlorofluorocarbons), HCFCs (hydrochloro-fluorocarbons), and the newer substitutes HFCs (hydrofluorocarbons). Until the mid-1970s, CFCs were used widely as spray-can propellants, solvents, cleaners and refrigerator coolants. But they were discovered to contribute to the breakdown of the ozone layer and, as a result, many countries signed the Montreal Protocol in 1987, which aimed to control the production and consumption of these gases. The combined tropospheric abundance of ozone-depleting gases peaked in 1994 and is now slowly declining. However, substitute HFCs, while less damaging to the ozone layer, still trap heat in the atmosphere and are adding to the greenhouse effect.

Since the ban on CFCs a gas called HFC-134a, or 1,1,1,2-tetra-fluoroethane, has been manufactured in ever growing quantities for use in air-conditioning systems. Scientists at the Norwegian Institute for Air Research reported in September 2006 that concentrations of the gas above Mount Zeppelin on the Arctic Island of Svalbard doubled between 2001 and 2004. Although halocarbons occur only in tiny amounts in the atmosphere, the warming effect they produce ranges from 3,000 to 13,000 times that of carbon dioxide. These gases do not break down easily and so can remain in the atmosphere for between five and one hundred years. Other synthetic compounds are even more potent: perfluoromethane (CF_4) lingers in the atmosphere for 50,000 years and sulphur hexafluoride (SF_6) is 22,200 times more effective a greenhouse gas than CO_2. Concentrations of both these gases are rising. Even if controls are brought in to further restrict their use, they will continue contributing to global warming for many centuries.

Water vapour

Water vapour (H_2O) is the most abundant greenhouse gas in the atmos-phere and the biggest contributor to the natural greenhouse effect. Concentrations vary spatially and are generally controlled by the Earth's temperature. The cold, dry air of the polar regions contains very little water vapour but the humid air over the tropics can contain up to four per cent H_2O. The "positive feedback effect" that a warmer world could induce is important; the warm air would absorb more water and therefore enhance water vapour's contribution to global warming (see box Runaway green-house on page 99).

Water is also a by-product of the breakdown of methane in the atmos-phere, so increased levels of methane may generate elevated concentrations of water vapour. Water vapour emitted from high-flying aircraft contributes to the formation of condensation trails, and these are linked to an increase in the formation of cirrus clouds; both contrails and cirrus clouds absorb heat and so help to warm the Earth's surface. At the present time the global budget of stratospheric water vapour is poorly understood and there are no reliable predictions of how concentrations may change in the future.

Cooling agents

Aerosols

Aerosols are tiny particles that float freely in the air. Natural sources include volcanoes, salt from sea spray, forest and grassland fires, and dust storms. Human activities, such as burning fossil fuels and altering land surfaces, also generate aerosols; around ten per cent of such particles in the atmosphere come from human activities, and these are concentrated in the northern hemisphere. The location of aerosols is important because they tend to cool areas above where they are concentrated. Incoming sunlight hits the surface of aerosols suspended in the atmosphere and is reflected directly back into space. This has the effect of reducing the amount of solar radiation reaching the Earth's surface. The effect has been called "global dimming".

The magnitude of cooling depends on the size and type of aerosol and the reflective properties of the underlying surface. Windblown dust and sea salt generate aerosols that are bigger than about one micrometre, while those smaller than one micrometre are generally the product of condensation processes. These include the conversion of sulphur dioxide gas (SO_2), such as that ejected by volcanic eruptions, to sulphate particles, and the formation of soot and smoke during burning. Once they are in the atmosphere, aerosols are mixed and carried by the wind and other climatic processes until removed, primarily during rainfall from clouds.

Aerosols play a crucial role in the formation of clouds, which also indirectly cools the planet. Small particles act as "seeds" upon which water condenses to form the droplets that cumulatively make up clouds. The concentration of aerosols dictates the type of cloud. Higher concentrations cause water in a cloud to condense on many separate particles, forming small droplets; clouds with smaller drops reflect more sunlight. Small particles fall slowly within the atmosphere, decreasing the amount of rainfall. This prolongs the life of the cloud. When concentrations of aerosols in a cloud are lower, the water condenses on fewer particles, forming bigger droplets that fall more readily as rain, reducing the life of the cloud.

Scientists have much to learn about the impacts of varying levels of aerosols in the atmosphere. Some believe it may be possible to offset future global warming by increasing the volume of aerosols in the atmosphere. A recent study by Tom Wigley of the National Center for Atmospheric Research in Boulder, Colorado, USA, suggests that injecting sulphates into the atmosphere, at the same time as cutting greenhouse gases, could prove more effective than either approach used alone. Wigley believes that injecting sulphate particles into the stratosphere, in similar amounts to that erupted by Mt Pinatubo in 1991, could provide a "grace period" of up to 20 years before major cutbacks in the greenhouse gas emissions would be required.

However, another piece of recent research suggests that, in some circumstances, aerosols can have a warming effect. Aerosols that absorb high amounts of solar radiation, particularly black carbon produced by incomplete combustion, tend to warm the atmosphere and inhibit cloud

HOW DIFFERENT ELEMENTS CONTRIBUTE TO EARTH'S TEMPERATURE

Radiative forcing is a measure of the influence a factor has in either warming or cooling the planet. The IPCC's 2007 report provides the following breakdown of elements contributing to our planet's elevated temperature, as a result of human activities. Values are for 2005 relative to the pre-industrial conditions of 1750 and are given in watts per square metre (Wm^{-2}).

• Carbon dioxide, methane and nitrous oxide: +2.3
• Aerosols produced by human activities (sulphate, organic carbon, black carbon, nitrate and dust): –0.5 plus –0.7 as a result of the energy reflected back from clouds formed as a result of aerosols
• Changes in tropospheric ozone due to emissions of ozone-forming chemicals (nitrogen oxides, carbon monoxide and hydrocarbons): +0.35
• Changes in halocarbons: +0.34
• Changes in albedo caused by land-cover change: –0.2 (the albedo is a measure of the amount of energy reflected back from the land surface. Pale surfaces, such as snow, have a high albedo; dark surfaces, such as forests, have a low albedo).
• Changes in albedo caused by aerosols of black carbon on snow: +0.1
• Changes in the output of energy from the sun: +0.12

The verdict
Human activities since 1750 have caused the planet to warm up, with a radiative forcing of +1.6 Wm^{-2}. Greenhouse gas concentrations alone would have caused more warming than observed because volcanic and human-made aerosols have offset some warming that would otherwise have taken place.

formation. This effect has been observed above the Indian Ocean and over the Amazon Basin.

Climate scientists versus the sceptics

As global warming has taken a grip on the planet, a small group of sceptics have argued that human-induced climate change is not happening at all. From the measurements noted at the start of this chapter, we certainly know that the global average temperature has risen by 0.76°C since records began, and that the planet is now warmer than it has been for some 12,000 years. Because cities absorb more heat than rural areas, the sceptics initially suggested that measurements taken near cities might exhibit exaggerated readings. However, new work suggests there is no bias. David Parker of the UK's Met Office analysed historical temperature records after dividing them into measurements taken in calm conditions and those taken during windy weather. If the sceptics were right, the readings taken in windy weather would be likely to be lower, given that the wind would disperse the extra heat. However, Parker's findings were that there was no difference between the two data sets. The IPCC FAR confirms this, saying: "Urban heat island effects are real but local, and have a negligible influence [on the world's temperature]".

The sceptics have also questioned how we know the extra warming is coming from anthropogenic rather than natural causes. Studies of Milankovitch cycles demonstrate that the warming we are witnessing does not correlate with orbital positioning that would be expected to warm the planet. However, there are other natural factors that influence the warming and cooling of the planet. For example, the output in energy from the sun is known to fluctuate, at times increasing the warmth reaching us and at times reducing it. Volcanic eruptions have a cooling effect, as they produce airborne particles that shade the planet for a year or two. It is now recognized that up to 40 per cent of the climatic variation since 1890 is probably due to a combination of solar cycles and the cooling effect of volcanoes. The sun's output is known to vary on timescales from decades to tens of thousands of years. A combination of low solar output and high volcanic activity is believed to have caused the Little Ice Age, a period of cooling that began around 1300 and ended around 1850. The 500-year cold snap brought bitterly cold winters to many parts of the world. Swiss glaciers advanced, crushing villages, Viking colonies in Greenland died out because they could no longer grow food, and Londoners skated on the frozen River Thames.

So could a change in solar output be responsible for the present global warming? In 2004, scientists published the results of a project that analysed records of trees preserved in riverbeds and bogs dating back 11,400 years. Like oxygen, carbon is composed of different isotopes. The most common is ^{12}C but there is also a very small fraction of ^{14}C. Records of past solar output are found in ^{14}C isotopes in tree rings and beryllium-10 (^{10}Be) in ice cores. Both isotopes are the products of cosmic rays striking the upper atmosphere. When the sun is more active, its solar winds magnetically fend off cosmic rays, reducing the quantities of ^{14}C and ^{10}Be produced. These fluctuations show up in wood because, as trees grow, they take up ^{14}C. Meanwhile, atmospheric records of ^{10}Be are captured in fresh snowfalls that over time become the ice from which scientists take cores. After correlating solar records with tree rings, which gave a close match, the researchers used the proxy to extrapolate the sun's energy output going back down the centuries. They discovered that there has been greater solar activity in the past 70 years than at any other time in the past 8,000 years. However, the study concluded that the current upsurge

GETTING THE MEASURE OF GLOBAL WARMING

A number of measurements are used to quantify greenhouse gas emissions.

• The content of greenhouse gases in the atmosphere is usually expressed in parts per million (ppm) by weight or parts per billion (ppb) by weight.

• The amount of fossil fuels we consume is generally referred to by the number of tonnes of carbon burned each year. At present, burning fossil fuels releases 7.2 gigatonnes (a gigatonne is a billion [metric] tonnes) of carbon into the atmosphere each year, in the form of carbon dioxide (CO_2). CO_2 weighs 44/12 times the weight of carbon. The atomic weight of the most abundant form of carbon is 12, while that of oxygen is 16. CO_2 is made up of one carbon and two oxygens, so the molecular weight of CO_2 is 12 + (2 x 16) = 44 and the molecular weight of carbon is 12. So CO_2 is 44/12 = 3.67 times heavier than carbon per molecule. Therefore, burning one tonne of carbon produces 3.67 tonnes of CO_2.

• It can be hard to envisage what a certain weight of CO_2 looks like. One kilogram of CO_2 at atmospheric pressure takes up 0.54 of a cubic metre, roughly equivalent to a human being and the space immediately around them. This equates approximately to the size of a coffin. Therefore one tonne of CO_2 is roughly the size of 1,000 coffins. This is the amount of CO_2 a single person is responsible for on one return flight from London to Egypt.

• Because greenhouse gases vary in potency and the length of time they linger in the atmosphere, a unit called global warming potential (GWP) is used to express the strength and lifespan of each gas in relation to CO_2. So CO_2 has a GWP of 1, methane's GWP is around 21, and that of nitrous oxide is 270. Multiplying the GWP of a gas by its prevalence gives the carbon equivalent, and enables researchers to consider all emissions as a group. Using the formula outlined above to multiply the carbon equivalent by 3.67 gives the CO_2 equivalent.

in solar output is not enough to account for the approximate 0.5°C temperature rise over the past 30 years.

Carbon isotopes are also helpful in demonstrating that the increased CO_2 in the atmosphere, like the rising temperature, has come from human rather than natural sources. The isotope ^{14}C is radioactive and has a half-life (the amount of time it takes for half of the atoms in a sample to decay) of around 5,700 years. Because fossil fuels are very old, the ^{14}C that was originally in them has already decayed. This means the CO_2 given off when we burn them has much less ^{14}C in it. The amount of ^{14}C in the air is therefore being diluted by our burning of fossils fuels. It is possible to

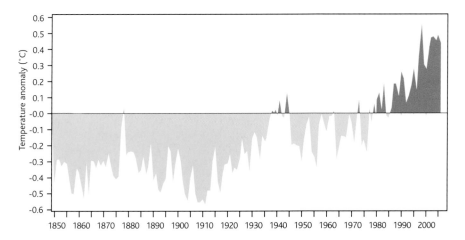

Graph showing the combined global land and sea surface temperature record from 1850 to 2006 shown as anomalies from 1961–1990 average.

estimate the change in ^{14}C in the air between 1850 and 1950 by measuring it in tree rings. When scientists calculate what this should be, based on human-sourced CO_2 emissions, the calculation matches the measurements well. This is proof that the rise in CO_2 is due to fossil fuel burning rather than a natural source.

Since the industrial revolution, we have been burning fossil fuels and clearing and burning forested land at an unprecedented rate. These processes convert organic carbon into CO_2. The 500 billion tonnes of carbon we have produced is enough to have raised the atmospheric concentration of CO_2 to more than 500ppm. The concentration has not reached that level because the ocean and the terrestrial biosphere absorb around 42 per cent of the CO_2 we produce. However, because we are adding CO_2 to the atmosphere faster than the ocean and biosphere can absorb it, the concentration of the gas is rising and this is pushing up the global average temperature. Measurements from ice cores show us that levels of CO_2 have played a strong role in regulating the temperature of the planet during its past switches into and out of ice ages. For example, a 3,270-metre core taken from Dome C in Antarctica shows that levels of CO_2 and methane marched closely in step with temperature over the last 650,000 years, a period spanning six ice ages and the warmer interglacials that divided them. The core also shows that today's level of 379ppm CO_2 is 27 per cent higher than previous naturally produced peaks of around 300ppm.

Doubters have also questioned whether the information derived from proxy climate indicators, such as ice cores and tree rings used by the IPCC in its assessment of climate, is an accurate representation of past conditions. In particular, they criticized the "hockey stick" curve produced by Michael Mann of the University of Virginia in 1998 using an amalgamation of data from ice cores, tree rings and coral. The graph showed the temperature variation over the past thousand years, which suggested a sharp rise in temperature caused by human activities. Climate scientists used the graph to claim that the 1990s were the warmest decade of the warmest century of the past millennium. The sceptics' main criticism was the reliance on "proxy data" to infer past temperatures. However, a report published in 2006 by a panel at the US-based National Research Council (NRC) largely vindicated the climate researchers' work.

The panel reviewed large-scale temperature reconstructions from different research groups to try and establish the Earth's surface temperature over the past 2,000 years. It reported that it had very high confidence that the last few decades of the twentieth century were warmer than any comparable period in the past 400 years. Although it said it was unable to confirm the original conclusion of Professor Mann's work, that the 1990s were the hottest decade and that 1998 was the hottest year, because of the difficulties in estimating the past climate over short timescales, it concluded: "Based on the analyses presented in the original papers by Mann et al and this newer supporting evidence, the committee finds it plausible that the northern hemisphere was warmer during the last few decades of the twentieth century than during any comparable period over the preceding millennium."

The environmental journalist and author George Monbiot investigated the motivation behind the sceptics' continued peddling of disproved theories in his book *Heat: How to stop the planet burning*. In researching the book, Monbiot uncovered an "active campaign of dissuasion" fuelled by funding from the tobacco company Philip Morris and the oil company ExxonMobil. For example, he found that Exxon funds a wide range of organizations all of which encourage doubt about the existence of climate change by selectively disseminating scientific information to suit its own agenda. Many have names that make them look like grassroots citizens' organizations or academic bodies; for example, the Center for the Study of Carbon Dioxide and Global Change, the National Wetlands Coalition and the National Environmental Policy Institute. As the world's most profitable organization (Exxon's net profit for 2006 was US$39.56bn, the highest ever reported by a US business) that makes much of its money from oil, it has more to lose than any other company from efforts to tackle climate change by reducing the use of fossil fuels.

Rather than being sceptics, Monbiot believes the people involved in perpetuating doubt about climate change to be "members of a public relations industry that begins with a conclusion and then devises arguments to support it". The result of this campaign has been to delay the global action on climate change to the extent that it may be too late to avoid some drastic changes to our planet.

CHAPTER TWO

How the world works

THE PROCESSES THAT DRIVE OUR GLOBAL CLIMATE AND HOW HUMAN ACTIVITY IS CHANGING THEM

Scientists believe that climate change is increasing the intensity of hurricanes.

To understand how unnaturally high levels of greenhouse gases can have a global effect on climate, it is necessary to know a little about how the world works. It is the sun that drives Earth's climate machine. As solar energy radiates into the atmosphere, some strikes dust particles or clouds and bounces back into space and some is reflected back from the surface of the land or oceans. The rest is absorbed by the Earth and then radiated back into space and the atmosphere. Every part of the Earth's climate machine

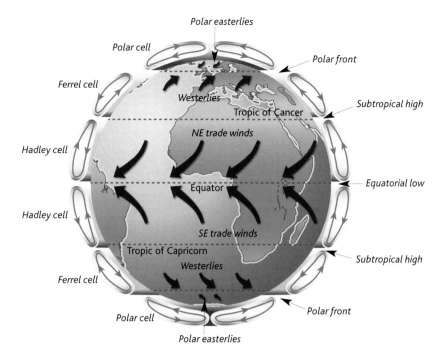

gains some energy each day through absorption and loses some through radiation and reflection. Apart from at latitudes of 40 degrees south and north of the equator, where this heat budget balances, the land either heats up or cools down.

The angle of the sun controls how much energy different parts of the Earth receive. Close to the equator, where the sun is high in the sky, the land and sea absorb plenty of incoming radiation and so temperatures rise. At the poles, where the sun strikes the surface more obliquely and the snow reflects back much of the incoming radiation, the temperature falls. If no processes existed to redistribute this heat, the poles would grow colder each year and the equator hotter. However, winds and ocean currents prevent this from happening by carrying warmth from the equator towards the poles.

Global winds

The trade winds are one such heat-carrying mechanism. In simple terms, hot air rising at the equator creates an area of low pressure and, as a result, air moves in from higher latitudes to fill the void. Compensating air-flows move air in the uppermost troposphere above the trades, away from the Earth's circumference. As this air cools on its path towards the poles, it sinks at latitudes of between 20° and 30° latitude before most of it returns to the equator.

If the Earth were not spinning, the wind would flow directly towards the poles, but because of the spin it is deflected right in the northern hemisphere and left to the south of the equator. This is an effect known as the Coriolis force. It creates the north-easterly trade winds in the northern

hemisphere and south-easterly trade winds in the southern hemisphere. In past times the trade winds were considered so reliable that sailing clippers regularly used them to carry goods around the world. The trades converge at the equator at the Inter-Tropical Convergence Zone (ITCZ), also known as the doldrums. The rising air of this zone creates low pressure systems that are associated with heavy rain storms and weak winds that can maroon sailors for days, hence the phrase "being in the doldrums".

The effect of the Earth's tilt is to make the sun appear to move north and south between the Tropics of Cancer and Capricorn as the seasons pass. The ITCZ does not stay in the same place but shifts back and forth across the equator following the sun's zenith. In the Indian Ocean, north-east trade winds blow throughout the northern hemisphere's winter months. But in summer, when the midday sun is overhead at the Tropic of Cancer at latitude 23.5°N, the ITCZ shifts north of the equator. The south-east trade winds now cross the equator, and are deflected to the right by the Coriolis force, backing up to become the south-west monsoons. This altered air-flow picks up moisture as it crosses the Indian Ocean and then unleashes heavy seasonal rains on India and South-east Asia.

Although most of the sinking air at around 30° latitude returns to the equator, some moves towards the poles. This moving air is also deflected by the Coriolis force, forming the "prevailing westerlies" that in the northern hemisphere dominate weather across northern Europe, North America and Canada. These westerlies are greatly affected by travelling cyclones and anticyclones that cause their actual direction to fluctuate from day to day. In the southern hemisphere, owing to the absence of landmasses, the westerlies blow almost continuously around the Earth. They are known as the Roaring Forties.

Cold, dense air also moves out from the poles forming the polar easterlies. Where this air meets the warmer air of the westerlies, the colder air pushes the warmer air upwards. This produces a zone of cyclonic low pressure where frontal depressions form. Some air then travels back to 30 degree latitudes, where it sinks and is recycled north and some moves back to the poles. Essentially, then, there are three major convection cells between the equator and each pole. They are known as the Hadley cell, Ferrel cell and Polar cell.

High up in the atmosphere near the tropopause, jet streams form at the four major boundaries where warm and cool air masses meet. These are meandering bands of high-speed winds that are many thousands of kilometres long and up to a few hundred kilometres wide but only a few kilometres deep. Their speed ranges from 100 to 200 knots. The sub-tropical jet forms at around 30° latitudes and the polar jet at around 60° latitudes. The jet streams take a meandering route as a result of Rossby waves. These form when air is forced to ascend on encountering barriers such as mountain chains. In the northern hemisphere, air that is forced to rise turns to the left, and as it descends again, it tends to turn to the right, inducing a ridge and trough pattern to the westerlies. If the ridge and trough loops become very pronounced, they detach masses of cold or warm air. These become the cyclones and anticyclones that are responsible for day-to-day weather patterns at mid-latitudes.

Birthplace of hurricanes

The low pressure of the Inter-Tropical Convergence Zone is also where hurricanes form. Associated with clusters of rain or thunderstorms, hurricanes are inwardly spinning weather systems that suck up warmth and moisture from tropical and sub tropical oceans and then carry the heat away from the equator. The term hurricane, which is used to describe such events happening in the North Atlantic, comes from Huracan, the name of a Mayan god of storms. In South-east Asia and the Pacific they are called typhoons, and in the Indian Ocean and Australia they are known as cyclones. They form when sea-surface temperatures exceed 27°C and the surrounding atmosphere is calm. In the northern hemisphere the hurricane season runs from June to November and in the southern hemisphere from October to May.

Warm air initially rises from the sea surface, creating a region of low pressure which draws in air from outside. The wind lifts moisture-laden air from the ocean, which helps swell the storm and so the process continues. At wind speeds of below 37 kilometres per hour (kmph), the storms are known as tropical depressions. They are steered in part by the Earth's rotation and in part by wind and weather systems. Tropical cyclones in the northern hemisphere commonly move west before turning polewards and then being blown east by the trade winds. Cyclones in the southern hemisphere are generally deflected south.

Evolving weather systems are set spinning by the Coriolis effect once they reach 10° latitude. They spin anticlockwise in the northern hemisphere and clockwise in the southern hemisphere. As wind speeds increase to 63kmph, a tropical depression is upgraded to a tropical storm. If the wind reaches sustained speeds of 120kmph, it officially becomes a category 1 hurricane. Category two grading requires sustained wind speeds of 154kmph; category three, four and five demand sustained wind speeds of 179kmph, 210kmph and 250kmph respectively. A single storm can be between 200 and 2,000 kilometres across and over its lifetime can release as much energy as a million Hiroshima nuclear bombs.

Signs of change

In 2005, the USA experienced its most destructive natural disaster when Hurricane Katrina hit the Gulf of Mexico. The hurricane had sustained winds of 225kmph, which generated a storm surge of seawater several metres high and flooded 80 per cent of New Orleans. It was the storm of the century... or was it?

In the past three and a half decades there has been a large global increase in the number of strong hurricanes. Since the 1970s, the number of category 4 and 5 storms has almost doubled from 20 per cent to 35 per cent. In 2005, there were a record 26 tropical storms and hurricanes in the North Atlantic, three of which reached category 5 strength. The season was unusual in other ways, too. Hurricane Wilma was the most intense hurricane ever recorded in the North Atlantic, while the cyclone that battered the coasts of Guatemala and El Salvador in May was the first ever to make landfall from the Pacific. Then there was hurricane Catarina, the first ever known to have come from the southern Atlantic. This ocean was previously thought to be too cold for hurricanes to form.

Scientists believe there is a link between global warming and changes to hurricane patterns. Researchers based at the National Center for Atmospheric Research, Colorado, USA, analysed sea-surface temperatures from around the world since the early twentieth century and noted that the tropical Atlantic between 10 and 20 degrees north was 0.9°C above the 1901–1970 average during the 2005 hurricane season. Although natural cycles such as El Niño (see page 60) and year-to-year variability did contribute to the greater sea-surface temperatures, they concluded that global warming was the cause of around 0.45°C of the abnormal warmth. The 125 experts who attended a workshop hosted by the World Meteorological Organization in December 2006 agreed that "given the consistency between high resolution global models, regional hurricane models and 'maximum potential intensity' theories, it is likely that some increase in tropical cyclone intensity will occur if the climate continues to warm". The IPCC went further in its 2007 Fourth Assessment Report (FAR), saying "There is observational evidence for an increase of intense tropical activity in the North Atlantic since about 1970, correlated with increases of tropical sea-surface temperatures".

Driving the oceans

The general circulation of the wind influences the movement of water in the oceans. The winds drive large surface currents in the top few hundred to a thousand metres, such as the Antarctic Circumpolar, Peru and Kuroshio Currents. These currents flow in large rotating loops called gyres. The Coriolis force sends the gyres spinning clockwise in the northern hemisphere and counterclockwise in the southern hemisphere. Like the winds, the world's ocean currents help to distribute heat around the planet.

There is also an element of vertical circulation, which is driven by a combination of temperature and salinity. Salty water is denser than freshwater, and cold water is denser than warm water. As the Gulf Stream carries warm, salty water up from the tropics into the Atlantic, the water cools and becomes denser. An offshoot called the North Atlantic Drift continues to around 80°N, where it eventually becomes so cold and dense that it sinks and then returns to the southern hemisphere at depth. The deep returning current is known as the North Atlantic Deep Water (NADW). This oceanic conveyor was originally called the thermohaline circulation (THC), but the term Meridional Overturning Circulation (MOC) is used more frequently today. It shifts water in a continuous flow around the globe.

Owing to the heat released to the atmosphere by the MOC, countries of north-west Europe are warmer than countries at the same latitude elsewhere. Scientists estimate that the North Atlantic part of the global conveyor carries a quarter of all heat transported from the equator to the poles. If the speed of the MOC decreases, it could therefore have an impact on climate around the globe. As Chapter 1 explained, climate change scientists have learned that the MOC stopped in the past when pulses of freshwater rapidly diluted the north Atlantic. A prime concern is that, if the planet warms sufficiently to melt major ice sheets, the less dense freshwater added to the ocean at high latitudes could prevent the MOC from overturning, thus bringing the warmth-giving oceanic conveyor to a halt.

Signs of change

In 2004, scientists at the National Oceanography Centre, Southampton (NOCS), placed a line of flow-measuring instruments across the Atlantic at 25° latitude, from the west coast of Africa in the east to the Bahamas in the west. The researchers then analysed measurements of temperature and salinity taken from research ships in the Atlantic in 1957, 1981, 1992 and 1998, together with the records from the instrument array.

They found that the MOC had slowed by around 30 per cent between 1957 and 2004. While the flow of the Gulf Stream northwards at 25°N had remained nearly constant, southward moving waters had become more sluggish. The NADW has two components: an upper current at depths of between 1,000 and 3,000 metres that originates in the Labrador Sea, and a lower current that flows at below 3,000 metres depth and originates in the Greenland-Iceland-Norwegian Sea. The researchers found that the upper current had remained reasonably constant, varying between 9 and 12 sverdrups (a sverdrup [Sv] is a unit measure of volume transport, with 1Sv equivalent to 106 cubic metres per second). However, the lower NADW transport had decreased by almost 50 per cent from 15Sv in 1957 to 7Sv in 2004. They also noted that the 2004 measurements showed more of the northward Gulf Stream flow was re-circulating back southward within the subtropical gyre and less was returning southward at depth.

Continued examination of the current's activity revealed that, in November 2004, the flow stopped completely for ten days. However, scientists do not yet know whether this is part of the current's natural variation or an abnormal event. Work is ongoing to assess the likelihood that global warming could cause the MOC to come to a complete halt. If the current remains as weak as the 2004 measurements, the UK's temperature could drop by 1°C over the next decade. A complete shutdown could lead to a 4°C to 6°C cooling over 20 years. The FAR predicts that the MOC will very likely weaken by around 25 per cent during the twenty-first century, but suggests temperatures in the Atlantic will increase despite such changes, owing to the much larger warming associated with forecast

Opposite: In the North Atlantic, warm water flowing up from the tropics keeps the countries of northwest Europe warmer than countries at the same latitude elsewhere.

Below: The global circulation system of the world's oceans.

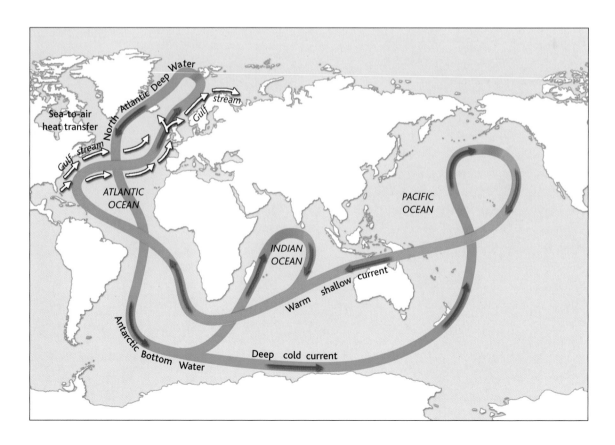

increases in greenhouse gases. However, the models used to simulate the current do not include potential changes in the rate of thawing of the Greenland ice sheet, which are difficult to predict. A melting of the ice sheet during 1,000 years would imply an average meltwater influx of around 0.1Sv, similar to that estimated for the Heinrich events explained in Chapter 1. It is likely, then, that future evolution of the Atlantic MOC is closely tied to the fate of the Greenland ice sheet.

Ocean climate cycles

The North Atlantic Oscillation (NAO) is a climatic cycle that operates independently from the ocean. It is driven by the change in air pressure between the high that lingers over the Azores at latitude 40°N and the low located above Greenland at around latitude 60°N. If both are at their extremes, the NAO is said to be positive, and the gradient strengthens the west-to-east jet stream's path across the Atlantic. The strong westerly winds bring mild, damp winter weather to northern Europe but also push more storms onto land. Meanwhile, southern Europe enjoys dry, warm conditions. When the air pressure rises over Greenland and falls over the Azores, the resulting negative NAO blocks the jet stream, and lets in cold, dry air from Asia to northern Europe. Southern Europe tends to be wetter during this phase.

The NAO is responsible for the variability in winter climate across the North Atlantic region, ranging from central North America to Europe and into northern Asia. The index varies from year to year, but tends to remain in one phase for intervals lasting several years. The missionary Hans Egede Saabye kept a diary in Greenland between 1770 and 1778, in which he noted; "In Greenland, all winters are severe, yet they are not alike. The Danes have noticed that when the winter in Denmark was severe, as we perceive it, the winter in Greenland in its manner was mild, and conversely". The winters of 1917, 1936, 1963, and 1969 experienced negative NAOs, but since the late 1970s the NAO has been increasingly positive. Of the 21-year period between 1981 and 2001, the winter NAO was positive for 15 years and negative for only six. Researchers are as yet uncertain whether this is a result of natural variability or whether the cycle has been influenced by anthropogenic climate change.

A similar climate cycle involving oceanic and atmospheric processes is the El Niño/Southern Oscillation of the Pacific Ocean. Here, at tropical latitudes, the trade winds blow consistently west, driving the surface waters towards South-east Asia. The current causes the sea level to be as much as 50 centimetres higher in the western Pacific than in the east. The ocean water here, close to Indonesia, is also remarkably warm; water in the western Pacific Warm Pool can be 30–32°C. On the other side of the Pacific, close to Ecuador and Peru, the water is much cooler as the offshore currents allow cool, deep water to upwell. Under normal circumstances, the warm moist air of the west Pacific produces constant showers and thunderstorms that release high levels of rainwater on the islands below. The air that rises above the Warm Pool flows east at a high altitude, to later fall over the eastern Pacific. This suppresses rain and maintains dry climatic conditions along the west coast of South America. This east-west atmospheric circulation is named the Walker circulation.

In some years, however, El Niño acts to reverse this pattern of processes. El Niño means "the boy" in Spanish and was named in the 1500s by South American fishermen who noticed that a current of unusually warm water flowed along their shores every few years around Christmas. Over time the Warm Pool waters on the western side of the Pacific deepen until they extend to depths of around 200 metres. Then rain and thunderstorms appear, accompanied by uncharacteristic westerly winds, which help force a kink in the thermocline (the undersea boundary created where warm surface waters and cold deep waters meet). Warm waters flood across the Pacific until they reach the South American coast. This means the region of heavy rainfall also shifts east, producing drought conditions in western Pacific countries such as Indonesia and Australia and heavy rains further east in South America. Without the contrast in the east-west sea-surface temperatures to create a pressure change, the trade winds weaken and can sometimes even reverse, perpetuating the condition.

An El Niño phase ends when the thermocline kink hits the east end of the Pacific, causing a rebound that heads slowly west. Eventually, the temperature gradient returns to normal, the trade winds strengthen and cold waters once again upwell off South America. Sometimes, the "normal" conditions become even more pronounced, so that the easterly trade winds and oceanic circulation strengthen. This situation is called La Niña, or "the girl". The heavy rains and storms over the Warm Pool intensify and the cold upwelling strengthens off of the coast of South America. El Niño and La Niña generally alternate, separated by several years of normal conditions. Sometimes El Niño is short-lived but it can last for longer than a year and may take two or three years to subside completely. Although the primary impacts are felt in the tropical Pacific, a particularly strong or persistent El Niño can influence weather on a global scale. Sometimes the jet stream is disrupted, changing the locations and subsequent tracks of storms. Also, greater amounts of heat and moisture can be carried to high latitudes.

Signs of change

Studies carried out by the National Centre for Atmospheric Research in Boulder, Colorado, USA, suggest that global warming may be contributing to the increased frequency and severity of El Niño weather disasters. Its researchers concluded that El Niño's odd behaviour between 1991 and 1995 was caused by the influence of global warming on the Pacific. It noted that "both the recent trend for more El Niño events since 1976 and the

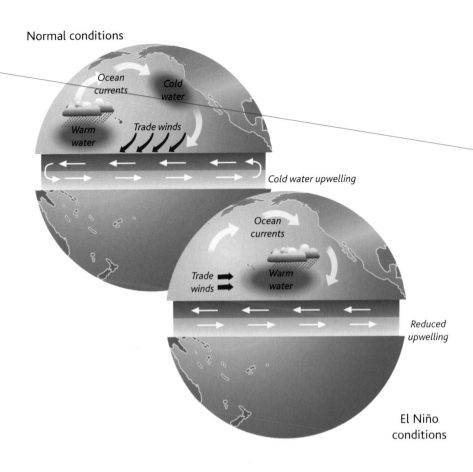

During an El Niño event, the trade winds push warm water east across the Pacific.

Normal conditions

Ocean currents

Cold water

Warm water

Trade winds

Cold water upwelling

Ocean currents

Trade winds

Warm water

Reduced upwelling

El Niño conditions

prolonged 1990–1995 El Niño are unexpected given the previous record, with a probability of occurrence of about once in every 2,000 years". It concluded that "…this opens up the possibility that the El Niño changes may be partly caused by the observed increase in greenhouse gases".

Other recent studies have shown that both the North Atlantic Oscillation and the ENSO simultaneously influence the summer Indian monsoon. Records dating back 132 years show that severe droughts wrought by a weak monsoon have always been accompanied by El Niño events, although El Niño events have not always produced severe droughts. Meanwhile, from 1870 to 1895, and from 1950 to 1995, strong Indian summer monsoon deluges were preceded by warmer than normal temperatures over Europe and North America in the previous winter and over western Asia in the previous spring, but colder temperatures over Tibet. The European temperature anomalies were related to the positive phase of the North Atlantic Oscillation (NAO).

Since the 1970s, the inverse relationship between El Niño and the Indian summer monsoon has weakened significantly. Over the same period, the NAO has strengthened. Scientists believe that the poleward shift of the jet stream over the North Atlantic and the resulting summer warmth experienced by northern Europe, have helped to counteract the effects of the warm ENSO. It is not clear whether the change in the NAO is related to global warming but it is clear that temperature is the

important link with the monsoon. It is also clear that the world in general is warming and over the past 50 years the largest annual and seasonal warmings have been at high latitudes.

It is possible that continued warming could bring more frequent warm weather to northern Europe, and this may, in turn, continue to strengthen the Indian summer monsoon. A much-increased monsoon caused by global warming would have serious consequences, as too much rain results in severe flooding and soil erosion, strongly affecting the population of that region. However, if the Meridional Overturning Circulation were to weaken and usher in colder conditions across the North Atlantic, the Indian summer monsoon would likely weaken. During times when Europe has experienced colder conditions, such as during the Little Ice Age between 1300 and 1850, the Indian monsoon has been weak. With more than half the world's population depending on the Asian monsoon to restock water supplies and bring much needed moisture for agriculture, this could be equally disastrous.

Under normal conditions, the west Pacific experiences high rainfall and thunderstorms. During an El Niño event, these shift east across the Pacific, bringing drought to Indonesia and Australia, and floods to South America. Here, cattle struggle to find dry land after floods strike Bolivia.

Oceans of acid

As well as distributing heat around the globe, the oceans play an important role in absorbing CO_2 from the atmosphere, as part of the carbon cycle. The oceans are naturally alkaline, with an average pH of 8.2, (pH is the scale on which acidity is measured and ranges from pH1, which is acid, through pH7, which is neutral, to pH14, which is alkaline). Carbon dioxide forms carbonic acid when it dissolves in water. In the natural cycle, photosynthetic plankton called coccolithophores take the carbon and use it to build their calcium carbonate shells. When they die they sink to the ocean floor, forming sediments that over millions of years become rocks, such as the chalk of the South Downs of England. If the level of acidity gradually increases, some of the calcium carbonate in ocean floor sediments dissolves, keeping the water at a constant pH level.

Signs of change

Many animals build their shells out of calcium carbonate, including coral polyps, lobsters and crabs. The calcium carbonate does not dissolve in the seawater because the water is saturated in carbonate ions; in other words, it simply cannot absorb any more. The recent rapid rises in greenhouse gases, however, are adding CO_2 to the water at such a rate the natural cycle cannot adjust sufficiently quickly. The result is that the seawater is becoming more acidic, so there are fewer carbonate ions and the creatures that build their shells from calcium carbonate are in danger of dissolving as the water tries to regain the natural equilibrium.

In 2005, the UK's Royal Society commissioned the first review of studies related to the acidity of the oceans. It concluded that the ocean is gradually becoming more acidic, having soaked up some 20–35 per cent of the CO_2 produced by our use of fossil fuels. If emissions continue at the present rate, the pH of the sea could fall by 0.5 pH units by the year 2100. No one really knows exactly how this will affect the marine ecosystem, but it is possible that the increased acidity will reduce the seawater's ability to absorb CO_2, making the ocean carbon "sink" less efficient, and leaving ever more of the greenhouse gas in the atmosphere. It is also possible that the increased acidity will hinder the ability of plankton to build their calcium carbonate plates, making the ocean's natural buffering system less effective.

Scientists are looking to the past to provide clues about how the oceans might respond. Sediment cores have revealed that, 50–55 million years ago, the Earth warmed up rapidly. The cause was not the regular Milankovitch cycles, but a 4,500-gigatonne injection of carbon into the atmosphere from undersea methane hydrates or volcanic eruptions. On reaching the atmosphere, the gas rapidly oxidized to CO_2. The resulting acidification of the ocean lasted for 100,000 years. Sea-surface temperatures rose by 4–5°C at the equator and at least 5°C in polar regions. While no major extinctions occurred on land, large numbers of deep-sea creatures were annihilated.

Tropical rainforests

The intense sunlight and rain-laden air close to the equator provide perfect growing conditions for many plants. As a result, the world's tropical rain-forests lie in a band around the planet's circumference, between the Tropics of Cancer and Capricorn. As the sun warms the moist air, it rises and

condenses to produce clouds, frequent thunderstorms and heavy rainfall. Temperatures stay relatively constant in the range of 22°C to 34°C all year round, while rainfall is in the region of 130 to 760 centimetres per year. Most of this falls during the summer rainy season between December and May. For example, the Amazon rainforest in Manaus, Brazil, receives 210 centimetres of rain per year, of which 30 centimetres fall in March, while less than five centimetres fall in August. The temperature here varies by only 2°C throughout the year. Fifty per cent of the rain in tropical rainforests comes from the trees recycling the water through evaporation, which creates more rain.

High levels of sunlight and rainfall make rainforests very biodiverse.

The largest rainforests are in South America, the Democratic Republic of Congo in Africa, and the islands of Indonesia. Other tropical rainforests lie in Hawaii and the Caribbean Islands. The Amazon rainforest is the world's largest, covering an area about the size of western Europe, or two-thirds the size of the continental USA.

Rainforests can be broadly divided into four horizontal sections: the forest floor, understorey, canopy and emergent layer. A whole range of ingenious adaptations enable plants to survive the high humidity and frequent dousings of rainwater. Trees often have splayed-out buttresses, which help to channel dissolved nutrients to the roots. Shrubs of the understorey and forest floor layers often have large leaves, which help them intercept the dappled sunlight. Many also have leaves with drip tips, which drain rainfall, promote transpiration and inhibit the growth of microbes and bryophytes that could damage or smother the leaf. Higher up, in the canopy, leaves are smaller as they are exposed to more sunlight and potentially damaging winds. Meanwhile, aquatic plants often have adaptations such as hairs that help them bob back to the surface if they are pushed under water. The plants that make up tropical rainforest ecosystems have been evolving in this way for 70–100 million years.

Tropical rainforests shift more carbon into and out of the atmosphere than any other terrestrial ecosystem. They are a vital carbon "sink", storing roughly 40 per cent of the carbon on Earth's continents. They are often described as "Earth's lungs" but in fact they may release little or no net oxygen. Although rainforest plants produce oxygen through photosynthesis, the ecosystem also consumes oxygen by respiration. In the past, tropical rainforests covered some 12 per cent of the planet's land surface (15 million square kilometres) but deforestation has reduced their coverage to less than five per cent of the terrain (6 million square kilometres). The clearance is ongoing; the Food and Agriculture Organization of the United Nations reports that between 2000 and 2005, 10.4 million hectares of tropical forest were permanently destroyed each year. This activity adds to the greenhouse effect in two ways. First, carbon stored in the plant biomass is released into the atmosphere as the trees are killed and, secondly, the ability of the rainforest to take up CO_2 through photosynthesis is reduced. Deforestation and fires in the Amazon account for more than 75 per cent of Brazil's greenhouse gas emissions and make it a significant contributor to global climate change.

Signs of change
In the past, scientists had believed that rainforests could help to soak up any excess CO_2 that humans injected into the atmosphere. Because plants use CO_2 to grow, it seemed feasible that any addition of the gas to the atmosphere would help the plants grow more, and that they would in turn become more efficient at removing CO_2. New work suggests, however, that the opposite is the case. After monitoring 18 virgin rainforest plots in the depths of the Amazon since the 1980s, tropical ecologist Bill Laurance of the Smithsonian Tropical Research Institute in Balboa, Panama, found that some of the 155 tree types were becoming more prevalent, and others less so. He and his colleagues found that two became significantly more common and 14 more rare over 15 years. The successful varieties were tall, relatively fast-growing canopy trees and the less successful were slower-growing trees living in the forest's shady under-layers.

Laurance and his colleagues could not find any past rainfall anomalies that might explain the change, and so reasoned that higher levels of CO_2 in the atmosphere could be the culprit. Greater concentrations of CO_2

speed up the process of photosynthesis, prompting trees to grow faster. Quicker growth favours the trees that already grow rapidly, so fast-growing species increase at the expense of slower-growing ones. Although scientists say it is hard to predict what the long-term changes to the rainforest might be, a higher concentration of fast-growing trees is likely to reduce the effectiveness of the tropical rainforest as a carbon store. This is because understorey trees grow more slowly and produce denser wood with a higher carbon content. A knock-on effect would be the impact on the animals living in the ecosystem. Many of the creatures are highly adapted to their environment, so any change could have a detrimental effect.

More recently, researchers have added to this gloomy outlook with the news that, when pollutants such as phosphorus and nitrogen taint rain-forest soils, they can trigger the release of CO_2. Cory Cleveland and Alan Townsend, from the University of Colorado, Boulder, USA, fed phosphorus fertilizer to a section of Costa Rica's tropical forest for two years. They discovered that the amount of CO_2 given off by the plants was 18 per cent higher than in the unfertilized control plots. Phosphorus and nitrogen produced by human activities are increasingly reaching rainforest ecosystems. For example, easterly winds carry significant amounts of phosphorus-containing dust from Africa to the Amazon basin, and desertification in the African Sahel region is enhancing this process. Levels of nitrogen and phosphorus in the air are rising because of the increasing use of fertilizers.

For tropical rainforests to thrive, they need vast quantities of water. At the heart of the Amazon lies the Amazon River which contains one-fifth of the Earth's freshwater and enters the sea in an estuary that is 320 kilometres across. Water is initially carried to the Amazon basin by the trade winds, which become laden with moisture as they travel over tropical waters. On reaching the Amazon, some vapour is released as rainfall. This is quickly recycled as the plants suck up water through their roots and release it back to the atmosphere through evaporation and transpiration from their leaves. This process of convection, which absorbs large amounts of solar energy, is repeated several times as the trade winds cross the continent. Scientists have shown that about half the rainfall of the entire Amazonian Basin results from water recycled by transpiration. It is one of the driving forces of the global climate and can influence the weather thousands of miles away. For example, poor rainfall in the Amazonian wet season can result in the USA experiencing drought during its growing season four months later.

The ENSO phenomenon has long been known to influence the amount of rainfall falling on the Amazon, with forests in the eastern Amazon experiencing drought during strong El Niño events. Usually a subsequent La Niña event serves to balance the water equation by bringing excess rain to the Amazon. In this way, the parched ecosystem is revitalized. For example, the El Niño events of 1997–1998 and 1982–83 brought droughts to the Amazon, but the following La Niña events brought anomalously high levels of rainfall, so the dry periods were relatively short-lived.

Rainforests need vast quantities of water to survive. The Amazon river contains a fifth of the Earth's freshwater. However, recent years have brought damaging droughts to the region.

Recent years, however, have brought more frequent El Niño events, without the rejuvenating La Niña process. The Amazon recovered little in 2004 following the drought of 2002–2003 and has since experienced another drought that began in 2005 and was ongoing when this book was being written in late 2006. Precipitation stayed below normal from 2002 to at least the end of 2005. The 2005 event resulted in a 16 per cent smaller sediment plume entering the Atlantic Ocean from Amazon run-off and 100 per cent more frequent forest fires across the basin.

The findings of a paper published in August 2006 suggest that the 2005 drought was caused by a warm tropical North Atlantic Ocean, which brought about a seesaw-like modification of the Hadley wind circulation cell. This resulted in rising air and abnormally high rainfall over the tropical North Atlantic, with subsiding air in the south over the Amazon and reduced rainfall across the western and southern Amazon. The same abnormally warm sea surface temperatures were responsible for the unusually strong hurricane season, which unleashed Katrina and other storms. As mentioned earlier, the frequencies of El Niño events and strong hurricanes have been forecast to increase with global warming.

Scientists are concerned that the forest may now be entering a critical point beyond which it is unable to recover. The warnings follow a study into the effects of drought on the Amazon, which found that tropical rainforests could not withstand more than three years of drought. Daniel Nepstad, an ecologist at Woods Hole Research Center in Massachusetts, erected 5,600 large plastic panels over a hectare of rainforest in order to divert rain and create an artificial drought. In the first year, photosynthesis slowed down to conserve water but the forest was resilient. When the lack of water extended into a second year, the plants sank their roots deeper to find moisture, but still survived. Eventually, some trees stopped growing. And four years after the start of the enforced drought they started dying. The tallest trees crashed down first, exposing the forest floor to the drying sun. By the end of the year the decaying trees had released more than two-thirds of the carbon dioxide they had stored during their lives. With drought having now affected the Amazon for two years, there are fears that trees could begin dying within a year. However, drought does not generally restrict rainfall to the 100 per cent reduction of the experiment. Also, rainfall deficits are quite variable regionally, so a forest-wide demise is, hopefully, unlikely.

Even if the forest manages to survive, however, the predictions are that climate change will bring more prolonged droughts to the region. And on top of the impacts of El Niño and abnormally warm Atlantic temperatures, deforestation is also taking its toll. Between May 2000 and August 2006, nearly 150,000 square kilometres of rainforest, an area larger than Greece, were destroyed in Brazil. Since 1970, over 600,000 square kilometres of the Amazon have been destroyed. Between 60 and 70 per cent of this was due to clearance for cattle ranches, with the rest resulting from destruction for small-scale subsistence farming. Logging results in forest degradation but rarely direct deforestation. However, studies have shown a close correlation between logging and future clearing for settlement and farming. Scientists have in the past predicted that the Amazon rainforest will not survive if more than 40 per cent of the biomass is lost. This is because the remaining forest will not be able to maintain the process of convection that recycles water throughout the ecosystem. Researchers using satellite images had estimated that the loss was around 17 per cent, but the images recorded total deforestation only where the canopy was removed. The figures did not include loss through selective logging of commercial plants snatched from beneath the canopy.

A study published in 2005 concluded that adding measurements of areas affected by selective logging in the Amazon doubled previous estimates of the total amount of forest degraded by human activities. If the 40 per cent mark is passed, and the new work suggests it is not far off, scientists believe that the world's largest forest will begin an irreversible process that will gradually turn it into desert. If they are right, the 90 billion tonnes of carbon released into the atmosphere by the dying forest will increase the rate of global warming by 50 per cent.

The world's drylands

The air that rises at the equator and moves towards the poles cools and sinks between latitudes 25° and 40°. This cool descending air stabilizes the atmosphere, preventing much cloud formation and rainfall. Consequently, many of the world's desert climates are found in this subtropical climate zone. In the northern hemisphere this latitude band contains the USA's Arizona Desert, large parts of Iran and Iraq, northwest India and Africa's Sahara. Southern hemisphere counterparts are South Africa's Karoo and Kalahari, Chile's Atacama and much of Australia. In these regions, temperatures peak at 45°C, although at night during colder times of the year, they can fall below freezing because of the high loss of radiation that results from the clear, cloud-free skies.

The Sahara is the world's largest desert, covering nine million square kilometres of northern Africa – 10 per cent of the continent – in dry sand and rocks. The recent discovery of fossils of the oldest known hominin *Sahelanthropus* in petrified dunes in Chad has dated the formation of the desert to more than seven million years ago. This potentially links its beginnings to the same past climate change that prompted expansion of the ice caps. The isthmus of Panama began to form at this time, which over millions of years diverted warming ocean currents sufficiently to encourage the ice caps' advance.

Scientists are aware that the desert has undergone changes in the past. Archaeological investigations show that, 22,000 to 10,500 years ago, the Sahara extended 400 kilometres farther south than today and was uninhabited apart from around the Nile Valley. Between 10,500 to 9,000 years ago, monsoon rains swept across the Sahara, transforming the area into a habitable land that was rapidly settled by Nile Valley dwellers. The continued life-giving rains, flourishing vegetation and animal migrations led to the rise of established settlements, in which farmers reared livestock.

Around 7,300 years ago, however, the monsoon rains retreated. As the desert dried up, Saharan communities withdrew once more to the Nile Valley. This withdrawal from the desert ultimately gave rise to the great Pharaonic civilization that emerged from 3500 BC. Within a few centuries, the Sahara had turned into one of the driest regions on Earth. Climatologists believe the change was prompted by the same phenomena that have dipped the world into and out of ice ages for millions of years: Milankovich cycles. Since around 9,000 years ago, a slow decrease in the tilt of the Earth's axis has resulted in cooler summers in the northern hemisphere. At the Sahara's latitude range, the initial change in the radiation was small, but researchers believe it was enough to weaken the monsoon over India and North Africa, thinning the vegetation. After a few thousand years, the stressed plants were no longer able to keep the soil moist and maintain the

cycle of evaporation, atmospheric circulation and precipitation needed to drive the African monsoon, and so the green Sahara rapidly reverted to the dry, barren state we know today. A model created by scientists at the Potsdam Institute for Climate Impact Research in Germany suggests the change could have happened in only 300 years.

Cacti growing in the Saguaro National Park in Arizona are well adapted for the dry conditions that prevail between 25° and 40° latitudes.

Signs of change

With the Sahara seemingly capable of switching rapidly between two climatic states, in a similar way to the Meridional Overturning Circulation, scientists are keen to know whether global warming could cause the desert to switch back to its green, vegetated state. A 2002 study, that looked at the vegetation cover across the southern boundary of the Sahara, found that increased rainfall had made farming viable in places that were previously arid and devoid of plants. The researchers discovered that vegetation had replaced sand across a great expanse of land stretching from Mauritania on the shores of the Atlantic to Eritrea 6,000 kilometres away on the Red Sea coast. The greening trend, which began in the mid-1980s, is believed to have resulted from increased rainfall. In one province, yields of sorghum and millet have risen by as much as 70 per cent in recent years.

On a less positive note, there have also been more floods along the southern edge of the Sahara. In 2003, heavy rains destroyed 180 ancient mud buildings, some more than 600 years old, at the UNESCO World Heritage Site in Timbuktu, Mali. The following year, officials declared a state of emergency in Gambia as the country suffered a massive locust infestation following unusually heavy spring rains. And in February 2006 torrential rain

Plagues of locusts have devoured crops in Gambia and Italy in recent years following bouts of unusual weather.

pounded the western Sahara desert, filling wadis and dry lake beds to overflowing and destroying refugee camps in south-western Algeria.

The greening does not seem to be reducing the size of the desert; rather it is shifting north as a result of changes to the Earth's global wind circulation patterns. In May 2006 scientists reported the results of analyses of temperature data for the lower atmosphere recorded by satellites over the past 25 years. They found that the air had warmed most at latitudes between 15° and 45° in both hemispheres. They concluded that this had widened the Hadley cell, that sees hot air rise at the equator to cool and subside at around 30° latitude, in turn causing the jet streams that form at the boundaries between the warm equatorial air and cool polar air masses to shift poleward, and expanding the width of the tropics by about 112 kilometres either side of the equator.

The implication is that the northern edge of the subtropics is also shifting poleward. If this is the case, we would expect to see below average rainfall and more frequent droughts affecting places that lie at around 40° to 45° latitudes, such as California and the Mediterranean in the northern hemisphere and southern Australia in the southern hemisphere. According to a report released in 2003 by the Department of Environment and Heritage of the Australian Government, average temperatures have risen across the country by 0.7°C in the last century. While rainfall has increased over the past 50 years throughout north-western Australia, it has decreased in south-western Western Australia, and in much of south-eastern Australia, especially in winter.

During the period 1952 to 1992, the number and frequency of heat waves affecting the Mediterranean has increased. Land records for the western Mediterranean show slight trends towards warmer and drier conditions over the last century. Surface water temperature records for the last 120 years show little overall trend but deep-water records for the western Mediterranean show a continuous warming trend since 1959. In 2005, the Mediterranean countries of Spain and Portugal suffered their worst droughts since the 1940s, while the heat wave in Algeria saw temperatures rise to 50°C. Freak swarms of locusts ravaged grape crops in Italy as its temperatures, too, climbed steeply.

Earth's extremities

As at 30° latitudes, sinking cold, dry air suppresses precipitation in the polar regions, giving rise to icy deserts. The southern polar region is occupied by the vast, mountainous continent of Antarctica, while the northern polar region encompasses the Arctic Ocean, continental Greenland, Alaska, parts of Canada and Siberia. Both areas have been covered in ice over considerable timescales because the little snow that accumulates rarely melts. The sun is not high enough to cause appreciable thawing and, historically, the temperatures have stayed mostly below freezing. During winter there is 24-hour darkness, which can give rise to extremely low temperatures. The lowest ever temperature recorded on Earth, –89.2°C, was at Vostok Station in Antarctica, during the winter of 1983. During the summer, which occurs between November and January in Antarctica and between June and August in the Arctic, both poles experience 24 hours of daylight. As a result of the perpetually chilly conditions, glaciers dominate the geography of the poles. There are two types of glacier: ice sheets, which flow out in all directions from a central point, and valley glaciers, which flow downhill from an upland source. The two main processes at work in glaciers are accumulation and ablation (melting). If the former happens at a rate faster than the latter, the glacier advances; if the reverse is true, the glacier retreats. The continental ice sheets of both Greenland and Antarctica have taken many millions of years to form.

Signs of change

It is the polar regions that are responding the most to human-induced global warming. The Arctic has been described as the "canary in the coal mine" of climate change, a reference to the way canaries were once used as warning devices to detect the presence of deadly gases in mines. Some scientists believe the ice is melting so fast that the Arctic Ocean will be ice-free during summer by 2060. Researchers at the US National Snow and Ice Data Center in the University of Boulder, Colorado, annually analyse satellite data to assess the conditions of the sea ice in September, as the Arctic's summer melting season draws to a close. After scrutinizing images from NASA, NOAA, and the US Department of Defense, together with

Canadian satellite data and weather observations, they concluded that the amount of sea ice present in 2006 was the second lowest ever. The lowest ever occurred the year before, when the volume of sea ice was 20 per cent lower than the average found from 1978 to 2001. The speed of melting is accelerating. Arctic sea ice melted at a rate of 6.5 per cent per decade between 1979 and 2001. That rose to 7.3 per cent after 2002, climbed to 8 per cent in 2005 and was at 8.6 per cent the following year.

In addition to the sea ice melting during summer, it now appears that less meltwater is refreezing in the winter. Scientists at NASA's Cryospheric Sciences Branch in Maryland, USA, found that in 2005 and 2006, the coverage of winter ice was six per cent smaller than the average over the past 26 years. As with the summer ice, the loss of winter ice appears to be accelerating; this retreat is much larger than the trend of 1.5 to 2.0 per cent decline in winter ice-cover observed per decade over the same period. And it is not just the sea ice that is disappearing. In 2001, NASA scientists concluded from a study of satellite images that the margins of the continental ice sheet were losing a metre in height each year. In 2004, they revised their measurement to ten metres per year, saying that in places the ice was dropping by one metre per month. A new study conducted by the University of Texas at Austin, USA, in 2006 found the ice sheet was melting three times faster than it had been five years before. The researchers concluded that the ice sheet was losing 200 cubic kilometres and contributing 0.5 millimetres to sea-level rise annually.

Glaciers and ice sheets are also dwindling on the other side of the world, in Antarctica. Although the continent receives just 2.5 to 7.5 centimetres of snowfall a year, a significant proportion of the world's freshwater – some say 70 per cent – is locked up in its ice. The average thickness of ice is several kilometres. Antarctica has two ice sheets, the East and West Antarctic ice sheets, which are divided by the Transantarctic Mountain Range. The 25 million-year-old East Sheet lies almost entirely above sea level, while the West Sheet, which is about one-fifth of the size of its eastern counterpart and much younger, lies mostly below sea level. Under the force of the weight of ice, large glaciers flow outwards, slowly moving ice from the centre of the ice sheets to the sea. When the ice reaches the sea and flows beyond it, it forms a large floating shelf of ice that is fixed to the continent. Such ice shelves cover 50 per cent of the Antarctic coast.

The findings of a study into ice around the Antarctic Peninsula, the most northerly part of the mainland, showed that in the past half century, 87 per cent of the 244 glaciers studied have retreated and that average rates of retreat have accelerated. Scientists from British Antarctic Survey and US Geological Survey drew this conclusion after analysing more than 2,000 aerial photographs dating from 1940, along with over 100 satellite images produced from the 1960s onwards, to work out the positions of glacier snouts. Some individual glaciers had retreated extensively. Sjogren Glacier, at the northern end of the peninsula, had moved back 13 kilometres since 1993, more than any other glacier in the study. Although 32 glaciers had gained ground, their advances were on average 300 metres per glacier. Subsequent studies of ice around the Amundsen Sea in West Antarctica have shown that the Pine Island and Thwaites Glaciers are now shedding more than 110 cubic kilometres of ice each year. This rate of discharge is triple what it was a decade ago. These two glaciers alone have the ability to raise sea levels worldwide by more than a metre.

Antarctica's ice sheets are also changing. The first study of gravity changes occurring within the continental ice confirmed in March 2006 that it is shrinking rapidly. The Gravity Recovery and Climate Experiment (GRACE) involved a pair of NASA/German satellites flying in formation 220 kilometres apart and measuring small changes in the Earth's gravity caused by shifts in the distribution of its mass. Scientists at the University of Colorado in Boulder, USA, who analysed the measurements found that Antarctica's continental ice shrank by around 150 cubic kilometres annually between April 2002 and August 2005. Most of this loss was from the West Antarctic ice sheet, and is thought to have contributed 0.4 millimetres to global sea-level rise each year.

The most startling change to Antarctica's ice came in 2002. Analysis of satellite images revealed that the northern section of the Larsen B ice shelf, a large floating ice mass on the eastern side of the Antarctic Peninsula, had shattered and broken away from the continent. Over 35 days, a 3,250-square-kilometre chunk of ice shelf disintegrated. During the preceding five years, the ice shelf had lost 5,700 square kilometres. It is now about 40 per cent of its original size. Six sediment cores, taken from the Antarctic Peninsula near Larsen B just before it collapsed, show the ice shelf had previously remained intact for 11,500 years.

In 2002, scientists discovered that a large section of the Larsen B ice shelf had broken away from the Antarctic continent. They blamed human-induced climate change.

Scientists at British Antarctic Survey believe human-induced climate change to be the culprit. They reported that stronger westerly winds in the northern Antarctic Peninsula were to blame for the marked regional summer warming that led to the retreat and collapse of the northern Larsen ice shelf. Analysis of weather balloons from the past 30 years shows that regional temperatures recorded in the mid-troposphere have increased at a rate of 0.5°C to 0.7°C per decade. This warming is three times larger than that observed globally. Air temperatures in the Antarctic Peninsula have risen by 2.5°C in the past half-century, five times faster than the global mean rate.

Pressure release

The pressure exerted by water and ice on the Earth's crust is considerable. A cubic metre of ice weighs some 0.9 tonnes; the same volume of water is slightly heavier, at one tonne. Over the past 650,000 years, sea level has risen and fallen some seven times by as much as 130 metres, as ice has repeatedly built up at the poles and then melted away. Evidence from the geological record suggests that, as the load on the underlying rocks shifted as ice ages came and went, the pressure change caused an increase in earthquakes and volcanic eruptions. For example, when warming ushered in the present interglacial period 10,000 years ago, the melting ice reduced pressure on the magma chamber beneath Iceland and prompted a burst of volcanic activity. Similarly, the release of land from the burden of ice triggered earthquakes around Scandinavia.

Scientists believe that the melting induced by climate change may be having the same effect. In 2004, NASA geophysicist Jeanne Sauber and geologist Bruce Molnia of the US Geological Survey linked the release of pressure on the crust from the melting of glaciers in southwest Alaska to an earthquake of 7.2 magnitude that shook the region in 1979. More recently – and of great concern – is an observed increase in the number of earthquakes beneath the melting Greenland ice sheet. When Göran Ekström and colleagues at Harvard University, USA, analysed the glacial seismic records back to 1993, they discovered that the number of quakes had risen in line with a number of sudden "slips" occurring within the ice. Between 1993 and 2002, the number of events ranged from six to 15 a year, but that jumped to 23 in 2004. During the first ten months of 2005 the number had already passed 30. Over 100 of the slippages originated in glacial valleys draining meltwater away from the main Greenland ice sheet.

The rise in sea level accompanying major ice melts can also trigger crustal movements and earthquakes. The redistribution of water following the last ice age is known to have prompted geological events around the rims of all the major ocean basins. The geological record contains evidence of 27 major landslides taking place in the north Atlantic basin over the past 15,000 years. Eight thousand years ago, shifts in the crustal pressures in the continental shelf off the west coast of Norway prompted an underwater

earthquake that triggered a massive slide of sediment off the edge of the shelf. Known as the Storregga collapse, the slide generated a tsunami more than 20 metres high in the Shetland Isles, off the north coast of Scotland, and up to six metres high along the east coast of the Scottish mainland. Although the region is now stable, a major quake close to Greenland could trigger the collapse of similar sediment piles.

Scientists predict that all the ice atop Mount Kilimanjaro in Tanzania will be gone by 2020.

Mountain glaciers

In addition to the polar ice there are glaciers that form under cold conditions at altitude. Even mountains in the tropics can have ice atop them; Mount Kenya and Kilimanjaro are examples. Other places exhibiting mountain glaciers range from the Cascades on the west coast of North America, to the Andes of South America, the Tien Shan mountains of northern China and Europe's Alps. The IPCC's Third Assessment Report defines mountain glaciers as one of the best natural indicators of warming, with the best ranking for reliability.

Signs of change
The World Glacier Monitoring Service was set up in 1986 to collect and publish data on glacier changes. Of the 88 glaciers it surveyed in 2002 and 2003, only four were growing and at least 79 were receding. In 2005, it reported that the mean annual loss in ice thickness of mountain glaciers had almost reached half a metre per year. This represents an overall loss of ice thickness of seven to eight metres since 1980.

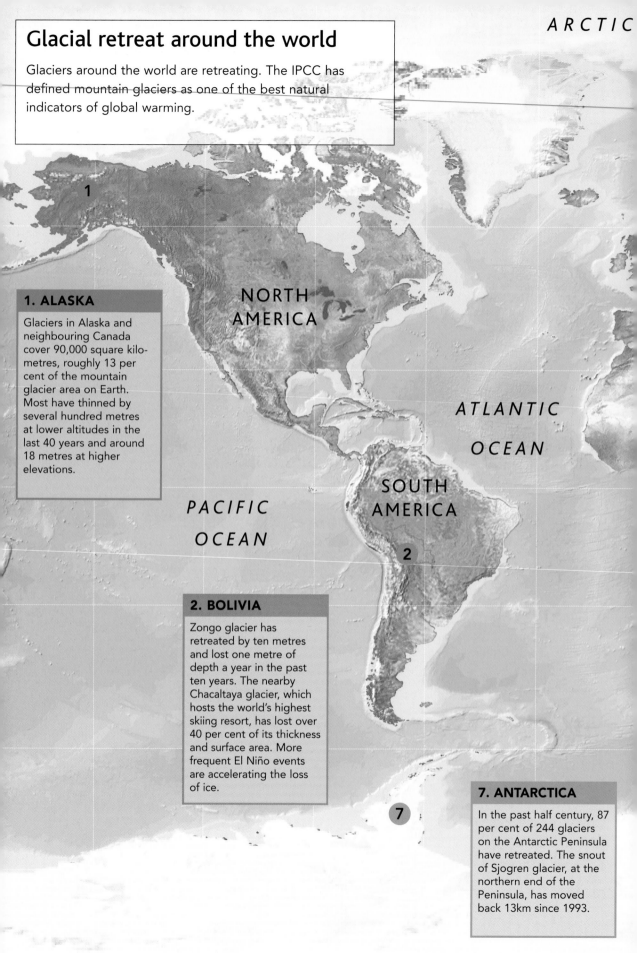

Glacial retreat around the world

Glaciers around the world are retreating. The IPCC has defined mountain glaciers as one of the best natural indicators of global warming.

NORTH AMERICA

ATLANTIC OCEAN

SOUTH AMERICA

PACIFIC OCEAN

1. ALASKA

Glaciers in Alaska and neighbouring Canada cover 90,000 square kilometres, roughly 13 per cent of the mountain glacier area on Earth. Most have thinned by several hundred metres at lower altitudes in the last 40 years and around 18 metres at higher elevations.

2. BOLIVIA

Zongo glacier has retreated by ten metres and lost one metre of depth a year in the past ten years. The nearby Chacaltaya glacier, which hosts the world's highest skiing resort, has lost over 40 per cent of its thickness and surface area. More frequent El Niño events are accelerating the loss of ice.

7. ANTARCTICA

In the past half century, 87 per cent of 244 glaciers on the Antarctic Peninsula have retreated. The snout of Sjogren glacier, at the northern end of the Peninsula, has moved back 13km since 1993.

OCEAN

6. SWITZERLAND

In the 1850s glaciers covered 4,474 square kilometres of the Alps. By the 1970s that was reduced to 2,903 and by the year 2000 it was down to 2,272 square metres. During 2004 and 2005, 84 of 91 glaciers in the Swiss Alps shrank. The Trift glacier lost 216 metres of its length over the course of a year. Europe's longest glacier, the 23-kilometre Aletsche in the southern Alps, lost 66 metres.

5. CHINA

Glaciers on the Tibetan Plateau, known as the Roof of the World, are shrinking at seven per cent annually. The Malan Glacier, which lies at the centre of the plateau, has retreated 45–60 metres over the past 100 years. In the 30 years from 1970, the Malan Glacier retreated 30–50 metres, at a rate of 1–1.7 metres per year. Baishui Glacier No.1 on Mount Yulong, the southernmost glacier of Eurasia, has decreased by 60 per cent from the Little Ice Age to the present.

6 EUROPE

ASIA

AFRICA

PACIFIC OCEAN

INDIAN OCEAN

OCEANIA

4. INDIA AND NEPAL

The Gangotri glacier, the source of the Ganges, India's holiest river, is receding by 23 metres each year. And the Khumbu glacier in Nepal, where Edmund Hillary and Tenzing Norgay began their acclaimed ascent of Everest, has lost five kilometres since they climbed the mountain in 1953.

3. TANZANIA

In 1912, maps show there were 12.1 square kilometres of ice on Mount Kilimanjaro. An aerial photograph shot in 2000 shows the ice to have retreated to 2.2 square kilometres; this means 80 per cent has disappeared. Scientists predict all the ice will be gone between 2015 and 2020.

SOUTHERN OCEAN

ANTARCTICA

Survey findings published in October 2006 from a study led by glaciologist Georg Kaser of the University of Innsbruck, Austria, show that loss to the world's glaciers and icecaps outside of Greenland and Antarctica has accelerated since 2001. Kaser's team combined different sets of measurements which used stakes and holes drilled into the ice to record the change in mass of more than 300 glaciers since the 1940s. They then extrapolated these results to cover thousands of smaller and remote glaciers not directly surveyed. They found that sufficient water drained from the world's glaciers and ice caps between 1961 and 1990 to raise global sea levels by 0.35–0.4 millimetre each year. For the period 2001 to 2004, the figure climbed to 0.8–1 millimetre annually.

Evidence of the meltdown is coming from all around the world. Kilimanjaro has lost 80 per cent of its ice since 1912; seven of Mount Kenya's glaciers have melted since 1900; a quarter of the ice in the Peruvian Andes has disappeared in the past 30 years; and Alpine glaciers have been retreating since Victorian times. In the Himalayas, which have the largest concentration of glaciers outside of the polar ice caps, covering 33,000 square kilometres, 67 per cent are in rapid retreat. Where glaciers are advancing, for example on Norway's west coast, increased snowfall appears to have overridden the effects of the warmer air temperature. But, in a news article published in *The Guardian*, Dr Kaser said that "99.99 per cent of all glaciers" were now shrinking.

Because many of the world's glaciers have been contracting for longer than human activity has been warming the planet, scientists believe there are causes other than human-induced global warming at play. A study of 169 glaciers by Hans Oerlesmans of Utrecht University in the Netherlands found that most glaciers reached a peak in the early nineteenth century before steadily retreating. An explanation is that the initial melting of glaciers was a response to natural warming since the Little Ice Age that occurred between 1300 and 1850. This should have tailed off during the twentieth century, but any slow-down has since been overtaken by the effects of human-induced climate change. In the Alps, melting had slowed to a halt by the mid-twentieth century, but since the 1980s it has accelerated rapidly.

Impacts on wildlife

Many species of plants and animals are highly adapted to the environments in which they live. Their size, shape, colour, feeding and sexual behaviours can all reflect the terrain, local weather conditions and biological characteristics of their surroundings. Changes in climate influence the size of populations, which in turn affects the distribution and abundance of species and, ultimately, the structure and function of the ecosystem. A long-term change in the global climate has the potential to dramatically change the distribution of species around the world. Those able to respond quickly or that can endure a wide range of conditions will likely thrive, but the more highly specialized species may simply die out as the conditions around them become intolerable.

There is a mounting body of evidence that climate change is already having an impact on wildlife around the globe. In 1999, the Royal Society for the Protection of Birds, World Conservation Monitoring Centre, English Nature and WWF jointly convened The Norwich Conference to collate information on wildlife changes observed by natural scientists and nature conservationists. The report *Impacts of Climate Change on Wildlife*, resulting from the gathering, concluded: "There is already clear evidence to show that wildlife from the poles to the tropics is being affected by climate change. Species migrations, extinctions and changes in populations, range and seasonal and reproductive behaviour are among a plethora of responses that have been recorded, and these are likely to continue apace as climate continues to change in the decades to come".

Signs of change

In the oceans, the increased temperature is having a noticeable effect. At the bottom of the marine food chain are plankton, microscopic creatures that are sucked up in their billions by marine animals such as Basking Sharks and whales. The Sir Alister Hardy Foundation for Ocean Science (SAHFOS) is an international charity that since 1931 has been collecting data from the North Atlantic and the North Sea on the biogeography and ecology of plankton. The findings of its Continuous Plankton Recorder (CPR) survey show that since the late 1990s there has been a progressive increase in the presence of warm-water or subtropical species. For example, the subtropical marine water flea *Penilia avirostris* has increased considerably in abundance in the North Sea in recent years. The Marine Climate Change Impacts Partnership reported in November 2006 that warmer-water plankton had shifted north by as much as 1,000 kilometres.

The CPR survey has also found that the ratio of the temperature-dependent copepod species, *Calanus helgolandicus* (warm temperature) and *C. finmarchicus* (cold temperature), has changed, with *Calanus helgolandicus* becoming more abundant and *C. finmarchicus* less so in recent years. The change has led to concerns that the ecosystem may get out of sync because warm-water plankton breed at a different time of year from the cold-water ones. Less cold-water plankton means that when cod larvae hatch, they have less access to their main food source. Overall, the *Calanus* biomass declined by 70 per cent between the 1960s and the 1990s. The observed changes may not be driven by climate change; abundance of *Calanus* correlates well with the North Atlantic Oscillation (NAO). However, it is possible that changes to the NAO are being driven by climate change. Either way, their continued decline is likely to have serious consequences for the North Sea's fishing industry. With fish stocks already depleted by overfishing, the lack of food for larvae could prompt the industry to decline or even collapse completely. In November 2006, the largest-ever analysis of global marine biodiversity and fisheries concluded that, for marine ecosystems to function, they need all their parts – from plankton to seabirds – intact.

Other changes, which may be linked to shifts in the distribution of plankton, have also been noted. Colin Speedie, an expert on the Basking Shark, who has been monitoring populations of the world's second-largest fish around the British Isles in recent years, has observed a marked rise in the numbers being spotted around Scotland and a drop in sightings to the south. In 2005, the Marine Conservation Society reported that sightings off the Scottish coast had increased by 65 per cent since 2001, while those off south-west England decreased by 66 per cent. Speedie believes that global warming may be prompting plankton to move north, with the Basking Sharks simply following their food supply. A single shark can filter a volume of seawater equivalent to that held in an Olympic-sized swimming pool, and can eat 10 to 25 kilos of plankton an hour.

While the Basking Sharks have headed north, the seas off of south-west England have begun to welcome more warm-water species. Ocean Sunfish, the world's heaviest bony fish, which can weigh up to 2,000 kilos, are being increasingly spotted by fishermen. They appear to have been lured to the area by the abundance of jellyfish and algae created by the warmer weather. Southern species, such as the Horse Mackerel and Gilthead Sea Bream are now being caught regularly off English shores, while warm-water species such as Triggerfish and Pufferfish have also been reported. A Slipper Lobster, normally found in the Mediterranean, was recently picked up off of Cornwall; only 20 of the creatures have been recorded in British waters since the mid-eighteenth century. While it is not certain that all these changes are the result of climate change, something is clearly driving warm-water species further north, and temperature is the likely culprit. The Marine Climate Change Impacts Partnership investigation found that sea-surface temperature (SST) and air temperature over the sea within the mid-latitude North Atlantic and UK coastal waters have been rising by 0.2–0.6°C per decade over the past 30 years.

The warmer conditions appear to be affecting breeding patterns, too. Winter atmospheric conditions over the North Atlantic influence the abundance of zooplankton eaten by Right Whales, one of the most endangered species of marine mammal. So-called because nineteenth-century whalers considered them the right whales to hunt, since they float when dead, there are three populations of them living today. Around 300 live in the North Atlantic, 8,000 live in the Southern Ocean and western South Atlantic, and an unknown number dwell in the North Pacific. Since 1971 scientists have built up a database of photo-ID images of a group of the Southern Right Whales during their annual gathering off Argentina's Peninsula Valdés between June and December. By studying individual females, the researchers created an annual index, charting the deviation of known whale births from the expected number of calves. Usually, a female gives birth every three years, but if a calf dies or is aborted, she needs a further two years to recover, so only reproduces every five years. The scientists charted the sea-surface temperature against the number of whale calves born for the years 1983 to 2000 and found that, as water temperature rose, calf output declined. The researchers believe that the decline is due to the warmer waters reducing the numbers of krill, the small crustaceans that whales feed on.

Although there is limited data on the diet of Southern Right Whales, and not enough direct evidence to tie krill concentrations in the western South Atlantic to the whales' reproductive success, other studies have found that sea-surface temperatures do affect krill abundance. Scientists from British Antarctic Survey compiled data from nine countries on the distributions of Antarctic krill in the first large-scale study of its kind in the area. They found that numbers of the species *Euphausia superba* had declined by about 80 per cent in the past three decades in the western South Atlantic. A localized reduction in the coverage of winter sea ice, associated with a temperature increase in the area, is the likely cause. At least half of the Southern Ocean's krill live in this sector of the Atlantic. The sea ice has algae growing on the underside, which is a food source for the krill; the ice acts as nursery ground for the juvenile krill in winter. The researchers found a relationship between the extent of the sea ice in winter and the numbers of krill the following summer.

Although the amount of sea ice has changed little across the whole of Antarctica since the late 1970s, the winter ice off the Antarctic Peninsula has dwindled. This is in line with a 2.5°C rise in temperature in the area, much higher than the global average. Declines in the populations of some krill-dependent predators have already been found in this region. As well as supporting whales, seals, penguins and other seabirds, krill are fished commercially. As *Euphausia superba* has declined, numbers of warmer water, jelly-like grazers known as salps, have invaded the higher latitudes. These are prey to some invertebrates, fish and seabirds but do not support any commercially fished species. A continued change in this ecosystem has implications both for fishing and wildlife tourism in the region.

Changes to the distribution of sea ice are also causing different species of penguin to shift their nesting sites. Historically, the Adélie Penguin, *Pygoscelis adeliae*, has lived among the sea ice that encircles the continent. Unlike all other species except the Emperor Penguin, *Aptenodytes forsteri*, it thrives only where the sea ice lingers well into spring. By contrast, the Chinstrap Penguin, *Pygoscelis antarctica*, breeds further north, away from the Antarctic circle. Together, these two very similar species form around 80 per cent of all seabirds in the Southern Ocean. Both birds require gravel nesting sites free from snow and meltwater, but the Adélie likes to be near sea covered by loose pack ice during the southern hemisphere breeding season, while the Chinstrap opts to be near ice-free waters. Since the last glacial maxima, some 19,000 years ago, Adélie Penguins have

How climate change is affecting wildlife

There is clear evidence from the tropics to the poles to show that wildlife is being affected by climate change. Changes have been recorded to populations, range and behaviour.

ARCTIC

5

5. ARCTIC

Polar Bears rely on drifting ice to hunt seals and are being affected by the diminishing pack ice. In Canada's Hudson Bay, the bears' population has dropped 22 per cent since the mid-1980s.

3

4

1

NORTH AMERICA

1. NORTH AMERICA

The range in the western USA of the Edith's Checkerspot Butterfly *Euphydryas editha* has shifted northward by 92 kilometres and upward in altitude by 124 metres over the twentieth century. The magnitude of this trend matches the observed warming trend over the same region. Sooty Shearwaters, once the most abundant summer seabird off the coast of California, declined by 90 per cent between 1987 and 1994.

ATLANTIC

OCEAN

2

SOUTH AMERICA

2. COSTA RICA

In 1987, 1,500 Golden Toads, a species found only in Monteverde, were recorded at the toad's principal known breeding site. But in 1988 and 1989 only one male was spotted. None has been seen since 1990, and the species is now probably globally extinct, the first documented casualty of human-induced global climate change. Since 1987, 20 out of 50 frog and toad species initially recorded in a demarcated study area have disappeared. The changes correlate with variations in mist frequency occurring in the cloud forest.

PACIFIC

OCEAN

8. ANTARCTICA

Since 1960, Adélie Penguins have shifted south to find new breeding grounds around the Ross Sea near loosened pack ice, as ice has disappeared from more northerly sites. Chinstrap Penguins, which prefer ice-free breeding locations, are meanwhile shifting southwards along the western coast of the Antarctic Peninsula. Air temperature records since the 1960s reveal a marked warming that is particularly evident in winter.

8

OCEAN

3. SCOTLAND

Since about 1950, the northern range margin of the Speckled Wood butterfly (*Pararge aegeria*) has shifted northwards by around 110 kilometres, corresponding with shifts in climate isotherms of 120 kilometres over the same period.

6. EUROPE

The range of the Sooty Copper butterfly *Heodes tityrus* is extending the northern limit of its range, while retracting the southern limit.

7. MALAYSIA

The United Nations Environment Programme says that the rising temperatures observed around the globe are having an impact on breeding as well as feeding in some species. Turtles are particularly vulnerable, because the sex of their offspring is dependent on temperature. Scientists have found that, at higher temperatures, some turtles are producing far more female eggs than male, and in parts of Malaysia, nesting sites are producing only females.

6 EUROPE

PACIFIC

OCEAN

AFRICA

7

INDIAN

OCEAN

OCEANIA

4. SPAIN

The chytrid fungus *Batrachochytrium dendrobatidis* has rendered the Midwife Toad *Alytes obstetricans* virtually extinct in Penalara National Park. There is a strong correlation between the decline in numbers and the rise in temperatures recorded between 1976 and 2002.

Many of the 130 Brown Bears that usually sleep out the cold season in Spain's northern mountains were still active during the winter of 2006 because milder weather ensured that they had enough nuts and berries to survive. Juan Carlos Garcia Cordon, geography professor at Cantabria University, said: "We cannot prove that non-hibernation is caused by global warming but everything points in that direction".

SOUTHERN OCEAN

ANTARCTICA

generally moved south and colonized areas that have opened up with the retreat of the ice sheets.

Over the past 50 years, however, Adélie populations have been disappearing from the northern part of their range on the west coast of the Antarctic Peninsula and islands such as the South Shetlands. This is in line with a marked warming that has caused the sea ice in this area to disappear at an accelerating pace. In the southern part of the Adélie's range, notably the Ross Sea, where the warming is loosening previously solid ice, breeding populations are on the rise. This is in direct contrast to the Chinstrap Penguin, which avoids pack ice and is often seen in open waters. The Chinstrap is instead extending its range southwards along the western coast of the Antarctic Peninsula in response to the disappearance of sea ice cover there. Increased warming is likely to see the losses to Adélie populations observed on the Antarctic Peninsula repeated further south, allowing the Chinstraps to set up breeding sites on the newly ice-free areas.

There is evidence that other communities of seabirds are suffering at the hand of climate change. A census of seabirds conducted seasonally off southern California, as part of the California Cooperative of Oceanic Fisheries Investigations programme, found a 40 per cent decrease in seabird numbers between 1987 and 1994. The continuation of low densities recorded during El Niño and La Niña events suggests that the observed impacts are due to long-term drivers, rather than yearly fluctuations in conditions. Sea level and ocean temperature off California increased drastically in the period 1949 to 1993, while the numberrs

Over the past half century, Adélie penguins have been moving to new areas with loose sea ice.

of zooplankton in the upper 200 metres of the water dropped by some 70 per cent.

The make-up of seabird communities off California also appears to be changing. Formerly dominant coastal and cold-water species, such as the Sooty Shearwater and Rhinoceros Auklet appear to be relinquishing territory to offshore and warm-water seabirds such as Leach's Storm Petrel and Black-vented Shearwaters. The Sooty Shearwater, a summer visitor which once numbered over five million birds and was the most abundant seabird off of California, declined by 90 per cent between 1987 and 1994. Because the bird is capable of covering vast distances, and breeds away from California, scientists are as yet uncertain as to whether its overall numbers are dropping or whether it is shifting its distribution to exploit new feeding grounds.

Marine species and seabirds that are relatively mobile may be able to colonize new favourable habitats with relative ease, but less mobile species are in danger of becoming trapped in increasingly unfavourable habitats. This is particularly the case with land animals, where urbanization and pollution have fragmented and degraded habitats to the extent that only isolated pockets of land with a suitable ecology remain. A study of frog populations in Costa Rica's Monteverde Cloud Forest Preserve suggests that climate change may have already claimed its first victim of extinction. Tropical montane cloud forests occur where moisture-laden trade winds blow onto a mountain range, leading to mist and cloud at high elevations. Often occurring as isolated habitat islands, they contain high concentrations of endemic creatures that have evolved to be intolerant of the surrounding lowland conditions. Studies of Monteverde's climate suggest that increases in tropical air and sea-surface temperatures since the 1970s have reduced the frequency of mist and shifted the humid zones, where clouds form, to higher elevations. The reduction in mist frequency has been accompanied by the disappearance of 20 of the original 50 species of frogs and toads from a study site following a particularly warm and dry period in 1987. That year, scientists counted 1,500 Golden Toads at the principal known breeding site, but only one male was seen there in 1998 and 1999. Since 1990, no one has spotted any. They believe the species, which was endemic to Monteverde, has become globally extinct.

There are many reports of population crashes in highland amphibian populations from around the world. A study published in early 2006 suggests that infectious diseases thriving under new temperature and water regimes have affected many harlequin frog species (*Atelopus*) across Central and South America. The disease-causing chytrid fungus *Batrachochytrium dendrobatidis* is believed to be responsible for the disappearance of about two-thirds of the 110 known harlequin frog species during the 1980s and 1990s. Amphibians' thin skin makes them highly sensitive to even minor changes in temperature, humidity, and air or water quality. It also makes frogs more susceptible to chytrid fungus. The study reported that temperature extremes may have previously helped to keep

PHENOLOGY: HOW NESTING BIRDS AND FLOWERING PLANTS HIGHLIGHT CHANGES TO THE CLIMATE

Phenology is the science of monitoring the times at which recurring seasonal events take place, for example the first appearance of migrating birds, the initial flowering of plants and the timing of when leaves begin to change colour and fall in autumn. So, phenologists in the UK record the times and locations at which bluebells first flower in spring and blackberry fruits ripen in autumn. In the USA they observe when Quaking Aspen trees first sprout leaves and Ruby-throated Hummingbirds arrive in North America after wintering in the tropics. And in South Africa they worriedly note the times of year when the Diamondback Moth begins to infest economically important crops of Oil-seed Rape. In Japan and China the time of blossoming of cherry and peach trees has long been associated with ancient festivals; some of these events are now being analysed by phenologists to glean information about the seasonal changes as far back as the eighth century.

The term "phenology" derives from a Greek word meaning an occurrence, or something that appears or comes into view. The discipline grew out of the Victorian vogue for monitoring and classifying the natural world. The eighteenth-century naturalist Gilbert White was an early phenologist, collecting information on the emergence of more than 400 plant and animal species between 1768 and 1793. Robert Marsham was another pioneer. A wealthy landowner, he kept systematic records of the "indications of spring" on his Norfolk estate from 1736. Consistent reports of events such as the first flowering or bud burst of plant species and the emergence or flight of insects were maintained by subsequent generations of his family until 1958. Between 1891 and 1948 a programme of phenological recording was set up across the British Isles by the Royal Meteorological Society. When the programme ceased, enthusiastic amateurs continued contributing records. In recent years, the UK Phenology Network run by the Woodland Trust and the Centre for Ecology and Hydrology has resumed phenological recording across the nation.

Because many of the changes recorded by phenologists are related to variations in climate, the discipline is now viewed as an important means of monitoring the effects of climate change. The USA has a national phenology programme; there is a European Phenology Network and other countries including Canada, China and Australia also have national programmes. Diligent recording by professional scientists and lay people alike is helping to show how climate change is affecting plants and animals. In August 2006, the journal *Global Change Biology* reported the findings of the world's largest study so far of seasonal events. Scientists from 17 countries analysed 125,000 records and observations made across 21 European countries between 1971 and 2000. The findings of 542 plant species and 19 animals showed that 78 per cent of all leafing, flowering and fruiting is now happening earlier in the year. They concluded that, on average, spring is arriving six to eight days earlier than in the past and advancing at a rate of 2.5 days per decade.

In the UK, the long timescale of records gathered by the Marsham family demonstrates that between 1850 and 1950 oak leafing dates became increasingly earlier. This correlates to a gradual climate warming. Another present-day phenologist, Jean Combes, collected oak leafing dates from 1947 and her reports show that leafing dates have continued to advance. Large trees appear to be reacting at different rates, with the sycamore, hawthorn and hornbeam coming into leaf noticeably earlier, and ash and common beech exhibiting smaller changes. This variation is likely to alter the competitive advantage of some species and lead to changes in the composition of woodland over the coming decades. The large leaves of the introduced sycamore may end up dominating woodlands because it has a tendency to shade out later-leafing native trees. Oaks may similarly come to dominate oak-ash woods because the oak appears to be adapting more quickly. Warmer

Phenologists record events in nature, such as the summer arrival of the Ruby-throated Hummingbirds in North America.

temperatures could also benefit the Small-leaved Lime, once widespread in woodlands across England, as it produces more fertile seeds under warmer conditions.

Some British plants, such as snowdrops and bluebells, prepare for the next season's growth in late summer and autumn, then sit out the winter as bulbs. In the past, this has given them a head start when spring comes around. By contrast, plants such as garlic mustard and cow parsley delay growing until the spring temperatures pick up. Warmer winters have enabled these plants to grow leaves much earlier than usual; as a result the species with bulbs have a reduced advantage. Some plants, including blackcurrants, rely on periods of winter chill to signal that they should break their dormancy. Winters with fewer cold spells may affect plants such as this by reducing their yield. As many fruits are grown commercially, this may have an economic impact.

Many plants and animals depend on each other for survival, so variations in response to climate change could disrupt important food chains. At the moment, most insects appear to be responding to the warmer temperatures at the same rate as their food plants. However, in one Oxfordshire wood there is evidence that the Blue Tit hatching season is no longer coinciding with the peak numbers of caterpillars in the woods. This may reduce the survival rate of the young birds.

A study of breeding birds in the UK published in 2005 by the British Trust for Ornithology and the Joint Nature Conservation Committee found that 33 species had, on average, begun laying their eggs up to 29 days earlier than 35 years ago. They include the Long-tailed Tit (laying 16 days earlier), the Greenfinch (13 days earlier) and the Oystercatcher (8 days earlier). Some migrant birds such as Tree Pipits, Reed Warblers and Swallows, which spend the winter months in warmer climes, are delaying their departure dates. And others, such as the Blackcap and Chiffchaff, are staying all winter, lured by the milder conditions.

the deadly disease in check, but new climate cycles are smoothing those annual temperature swings. Increased evaporation in parts of the tropical uplands has increased the number of clouds. This has decreased daytime temperatures by blocking sunlight, but has increased night-time highs by creating an insulating blanket. The change has created conditions in which the frog-killing fungus, which grows and reproduces best at temperatures between 17°C and 25°C, can thrive.

Problems arise for species when they are unable to shift their distributions any further south or north. A stark example is that of Polar Bears, which depend on sea ice to live, hunt and breed. With the Arctic warming at more than twice the rate of the rest of the world, sea ice is expected to disappear completely in summer before the end of the century. There are 19 Polar Bear populations in the world, amounting to some 20,000 to 25,000 individuals. They are spread across the Arctic in Russia, Greenland, Norway, Canada and Alaska. In 2001, only one population was in decline but, according to WWF, by 2006 this had risen to five. The two best studied of the dwindling groups are the western Hudson Bay population in Canada and the southern Beaufort Sea population which roam parts of Canada and Alaska. Their populations have declined by 22 per cent and 17 per cent respectively since the mid-1980s. Incidents of drowning Polar Bears, cannibalism and "problem bears" looking for food near to Inuit communities are becoming common, along with evidence of shifting distributions and changing diets. These observations are consistent with predicted changes caused by global warming. A report by 40 members of the Polar Bear specialist group of the World Conservation Union

The world's 19 populations of Polar Bears depend on sea ice to live, hunt and breed. Yet scientists predict all Arctic summer sea ice will be gone by the end of the century.

concluded in 2005 that the world's largest bear should be classified as a "vulnerable" species based on a likely 30 per cent decline in its worldwide population over the next 35 to 50 years. The panel stated that: "The principal cause of this decline is climatic warming and its consequent negative affects on the sea ice habitat of Polar Bears".

Our planet named Gaia

The impacts of climate change around the planet are proof that Earth is made up of a delicate balance of processes, and that altering one element can have far-reaching consequences. But realizing how the planet works has been a long, slow journey from the development of the Gaia Hypothesis in the 1970s.

The Gaia Hypothesis, developed by chemist James Lovelock and microbiologist Lynn Margulis in the 1970s, proposes that the interactions of air, ocean and soil constantly keep the Earth in a state most favourable for the plants and animals living on it. William Golding named the system and the hypothesis Gaia, after the Greek goddess of the Earth. Lovelock developed the idea in a book published in 1979 called *Gaia: A New look at Life on Earth*. In it he stated: "… the physical and chemical condition of the surface of the Earth, of the atmosphere, and of the oceans has been and is actively made fit and comfortable by the presence of life itself. This is in contrast to the conventional wisdom which held that life adapted to the planetary conditions as it and they evolved their separate ways."

The idea of Earth being a holistic living system has deep roots. According to traditional Aboriginal belief, all life on the planet today is part of one vast unchanging network of relationships that can be traced back to the spirit ancestors of the Dreamtime. Many native American groups have viewed the world in a similar way, as laid out in their verse "all my relations", in which the definition of relations encompasses family members, other two-legged beings, four-legged creatures, fish, birds, trees, rocks, thunder and "beings" under the earth. And the ancient Greeks, thinking this way, gave the Earth the name Gaia or Ge, as is recalled through the roots of the words geology and geography.

As the study of science evolved over time, however, the workings of the world began to be dissected to such an extent that the planet was no longer viewed holistically. The Victorians, masters of exploration, collection and classification, believed that life was explained as Darwinian evolution taking place against a dead, inorganic background. Only when astronauts got their first glimpse of Earth from outer space in the 1960s, and instruments began sending back the first scientific data from afar, did people realize that our planet stood out as a vibrant but fragile anomaly among the dead, lifeless planets lined up next to it in the solar system.

"For me, the personal revelation of Gaia came quite suddenly – like a flash of enlightenment", says Lovelock. "I was in a small room on the top floor of a building at the Jet Propulsion Laboratory in Pasadena, California.

It was the autumn of 1965 … and I was talking with a colleague, Dian Hitchcock, about a paper we were presenting … It was at that moment I glimpsed Gaia. An awesome thought came to me. The Earth's atmosphere was an extraordinary and unstable mixture of gases, yet I knew that it was constant in composition over quite long periods of time. Could it be that life on Earth not only made the atmosphere, but also regulated it – keeping it at a constant composition, and at a level favourable for organisms?"

An early criticism of Lovelock's hypothesis was that the self-regulating nature of Gaia implied that it was a teleological being, driven by a purpose or design. Another objection, from Darwinists, was that Gaia had evolved independently of natural selection. Lovelock responded to the criticisms by developing the Daisyworld model, an imaginary planet that maintains conditions for its survival by following its natural processes. Daisyworld revolves around the sun and absorbs energy at a rate dependent on the sun's luminosity and the albedo – or reflectivity – of the planet. It contains only two species of life: light daisies and dark daisies. The light daisies reflect radiation from the sun, thus cooling the planet, while the dark ones absorb energy, warming the planet. Lovelock used the dynamics of real daisy growth to create the model. How the different daisies grow depends on their present population, natural death rate, available space and temperature. When the model is run with the sun's luminosity gradually increasing, the populations of light and dark daisies adjust accordingly to keep the temperature at an optimal level for daisy growth. Thus, Daisyworld represents a viable ecosystem that regulates its temperature without relying on selection or teleology.

A real-world example is how marine algae help regulate climate. In winter in the Arctic, blooms of diatoms (microscopic algae with skeletons of silica) draw down carbon dioxide from the air and so cool the planet. Later in the ocean season, coccolithophores bloom and release some of the gas back to the air. They do this by changing calcium bicarbonate present in the ocean into the calcium carbonate they need to build their shells plus free carbon dioxide, which escapes to the air. Because coccolithophore blooms are white in colour, they increase the albedo (reflectivity) of the ocean and so have a cooling effect by reflecting more sunlight back into space. In warm tropical waters, the algae have another climatic effect because they are evenly distributed across the thermocline. Because of this the ocean is transparent and allows the sunlight to penetrate deep, helping to create an even distribution of solar energy across the ocean. Finally, nearly all algal species make a chemical called dimethyl-sulphinio-propionate to protect them from the saltiness of the sea. When they die this decomposes to dimethyl sulphide and some is carried in the atmosphere. Here, it provides condensation nuclei upon which clouds readily condense, and thus has a direct role in regulating climate.

Human-induced climate change now appears to be changing the natural balance of Gaia in a way that is not favourable for many of the

life-forms on the planet. So will the balance change in a way that might rid Gaia of her human irritants? Writing in his book *The Ages of Gaia: A Biography of our Living Earth*, Lovelock says:

> "Gaia emphasises most the significance of the individual organism. It is always from the action of individuals that powerful local, regional and global systems evolve. When the activity of an organism favours the environment as well as the organism itself, then its spread will be assisted; eventually the organism and the environmental change associated with it will become global in extent. The reverse is also true, and any species that adversely affects the environment is doomed; but life goes on. Does that apply to humans now? Are we doomed by our destruction of the natural world? Gaia is not purposefully anti-human, but so long as we continue to change the global environment against her preferences we encourage our replacement with a more environmentally seemly species."

James Lovelock, who developed the Gaia Hypothesis. This proposes that interactions between the air, ocean and soil keep the Earth in a state that is favourable to life forms living on it.

CHAPTER THREE

Predicting the outcomes of climate change

HOW OUR WORLD WILL LOOK IF WE DO NOTHING TO CUT GREENHOUSE GAS EMISSIONS

Governments face the challenge of predicting where drought and floods might strike in the future.

The Earth's ecosystems, from the icy poles to the Amazon rainforest and oceans, sustain human societies by providing water, food, medicines, timber and other resources. Climate change is already tinkering with many of these natural systems; the extent to which this will continue or accelerate in the coming years will depend on what efforts we make to reduce levels of greenhouse gases.

Without rapid and extensive mitigating action, scientists anticipate continued changes in "ocean circulation, sea level, the water cycle, carbon and nutrient cycles, air quality, the productivity and structure of natural ecosystems, the productivity of land used for growing crops, rearing live-stock and sustaining timber stocks, and the geographic distribution, behaviour, abundance and survival of plants and animals, including those that help to spread human disease". For governments to be in with a chance of preparing for the coming climate upheavals, they need to know how the planet will respond to the pressures placed on it by our penchant for hydrocarbons.

Clues from the past

As well as using contemporary measurements to observe recent changes, scientists have become adept at extracting clues on past climate fluctuations preserved in the natural world. Using these climate proxies, they have demonstrated that at various times in the past, Earth has been warmer and colder than at present. In fact, the Earth has shivered its way through at least 17 major ice advances in the last 1.6 million years, the most recent of which reached its peak some 20,000 to 18,000 years ago and came to an end about 10,000 years ago. They have shown that levels of greenhouse gases such as CO_2 and methane have helped regulate the temperature of the planet in the past, but that through deforestation and burning fossil fuels humans have injected such high levels of greenhouse gases into the atmosphere that they have overridden the Earth's natural ability to maintain its regular "interglacial" temperature. Here are some of the many methods scientists use to glean information on past climates.

Ice cores

Ice cores taken from glaciers and ice sheets contain some of the best records of past climates. The Greenland and Antarctic ice sheets have yielded some particularly valuable records because they are immensely thick and the ice has been relatively undisturbed. However, there are ice fields on six of the seven continents, so it is possible to study past climatic changes right around the world. Cores taken from high-altitude tropical and subtropical ice caps, such as Mount Kilimanjaro, the Andes of Peru and Bolivia and the Himalayan Plateau, have provided an invaluable insight into past climate fluctuations far from the polar regions.

Cores are usually around 20 centimetres wide and cut into lengths of a metre, to allow them to be easily catalogued and stored. While short cores of up to 40 metres are retrieved using a hand-powered auger, longer ones demand electro-mechanical drills. The longest core extracted so far is the Dome C core, taken from east Antarctica as part of the European Project for Ice Coring in Antarctica (EPICA), a ten-country consortium. Some 3.2 kilometres long, the core contains climate data going back 800,000 years.

As briefly explained in Chapter 1, isotopes are different atoms with the same chemical behaviour but different masses. They comprise a combination of protons, neutrons and electrons. The number of protons sets what the element is, but the number of neutrons and electrons can vary. Oxygen has eight protons and usually contains eight neutrons to form ^{16}O, but in some cases there are nine or ten neutrons. These give rise to ^{17}O and ^{18}O isotopes. In the same way, hydrogen occurs as ^{1}H (99.984 per cent) or ^{2}H (0.016 per cent). The different hydrogen and oxygen isotopes combine in various ways to produce molecules of water, H_2O. The two most used in palaeoclimatic research are $^{1}H^{2}H^{16}O$ and $^{1}H_2^{18}O$. Having oxygen and hydrogen atoms with extra neutrons, these form heavier molecules of water than the atoms with fewer neutrons.

When water is evaporated from the oceans, the lighter molecules with fewer neutrons evaporate more readily than the heavier ones. This means the vapour is enriched in lighter molecules and depleted in heavier ones. When the temperature of the air mass falls, prompting condensation, the heavier molecules condense out first. So the rain or snow that falls is initially enriched in the heavier molecules. If the temperature continues to fall, the condensation will contain decreasing concentrations of the heavy molecules. Ice sheets are particularly depleted in ^{18}O so, during ice ages, when large volumes of water become locked up in ice sheets, the oceans become progressively enriched in heavier oxygen molecules. The varying ratios of ^{18}O and ^{16}O present in ice cores directly reflect the changing temperatures of the air from which the snow fell.

All oxygen isotope concentrations are today represented as a deviation from a figure called the Standard Mean Ocean Water, developed in 1961. This means the results from different cores around the world can be easily compared. Scientists study the temperatures in conjunction with the thickness of individual ice layers in a core. In this way they can work out whether temperature variations were seasonal or part of much larger climatic variations taking place over thousands of years. Similar analyses, using deuterium isotopes, enable scientists to deduce varying levels of humidity.

Aside from isotopic analyses, scientists measure the levels of green-house gases such as carbon dioxide, methane and nitrous oxide. When snow falls, air bubbles remain trapped between the flakes and are preserved as the snow becomes slowly compacted into ice. If snow accumulates rapidly, the bubbles may be isolated from the atmosphere relatively quickly. For example, in Greenland this process takes around 100 years. But if accumulation is slow, such as in Central America, it can take 1,000 years for the bubbles to become cut off from external influences. This means the trapped bubbles invariably contain air that is younger than the ice in which they are trapped. However, they provide invaluable chemical snapshots of the atmosphere from close to the time the snow fell.

Scientists have measured concentrations of greenhouse gases preserved within cores to see how they changed during past transitions from ice age to interglacial and vice versa. The records may help us understand how our current climate could respond to the record levels of greenhouse gases. Measurements from EPICA's Dome C core show that there have been eight cycles of atmospheric change in the past 800,000 years, at which times carbon dioxide and methane both rose to peak levels. They also reveal that carbon dioxide concentrations always stood between 180 and 300 parts per million (ppm) during the last 800,000 years. Because levels have now risen to an unprecedented 379ppm, the scientists are limited as to how much they can learn from ice core records; in other words we are facing unknown territory. Methane concentrations never exceeded 750 parts per billion (ppb) before the industrial age, but are now at 1,780ppb.

Scientists also make deductions from levels of dust and particles present in ice cores. When explosive volcanic eruptions eject ash and sulphate particles into the air, these circulate the planet and often leave their signature as distinct visible layers. These can be useful in correlating ice cores extracted from different places. Similarly, layers of fallout from nuclear bomb tests conducted in the 1950s and '60s provide useful markers in time. Other forms of dust and chemical contaminants provide a variety of proxies. For example, chlorine and sodium ions found in Himalayan ice have been used to estimate the strength of the Indian monsoon. And salt trapped in the Dome C core has been used to calculate how far the sea ice extended around Antarctica every time the continent got colder. In South American cores, dust layers provide helpful indications of seasonal and climatic variations in wind strength.

There are limitations with ice cores, so scientists have to be extremely careful when interpreting the climatic information they contain. For example, when ice melts in summer, water can percolate down through cores and shift molecules around, thus contaminating the historical record. Also, cores only represent the conditions during periods of snowfall. If in the past there was no snowfall for a prolonged period, there may be a break in the record that puts the dating of annual layers out of kilter. Two cores are often extracted from nearby locations, so the layering can be compared and the quality of the data tested. Other problems arise if the ice becomes deformed close to the bedrock. For this reason scientists have to assess the bedrock conditions carefully before choosing where to extract a core. However, the Greenland and Antarctic ice sheets are so thick that it is not always necessary for scientists to recover ice all the way down to the underlying geology.

Sediment cores
Sediments continually build up at the bottom of lakes, seas and oceans. Pollen grains and charcoal fragments get washed by rivers or blown by the wind into the water, where they slowly sink to the bottom. Hundreds or thousands of years later, their preserved remains enable scientists to glimpse times when ancient wildfires ripped through forests, and see how

A scientist takes a sample of ice in Antarctica.

historic assemblages of plants vary from modern habitats.

Microscopic plankton that float in the water column or dwell on the seabed become incorporated into sediments when they die. Foraminifera shells are made of calcium carbonate ($CaCO_3$), while diatoms are composed of silicon dioxide (SiO_2). The ratios of stable isotopes bound up in their shells provide information on the water chemistry at the time they formed. And, just as stable isotopes in ice cores reveal past air temperatures, so those in sediment cores can tell us how warm the oceans used to be. Warmer waters tend to evaporate off greater amounts of the lighter isotopes, so shells built in warm waters become enriched in the heavier isotopes. Measurements of the isotopic ratios preserved within foraminifera and diatom shells have now been taken from hundreds of deepsea cores around the world. Scientists have used them to map past surface and bottom water temperatures.

An international team of scientists called the Arctic Coring Expedition (ACE) recently used ice-breaking ships and a drilling rig to take samples from beneath the Arctic seabed from Siberia to Greenland. They used the legacy of minerals and fossils in the sample cores to chart the changes that took place in the Arctic around 50–55 million years ago. This is when the Arctic suddenly warmed in a naturally occurring event called the Palaeocene/Eocene Thermal Maximum. The scientists discovered that, 55 million years ago, Arctic summertime surface-ocean temperatures were comparable to those today of the sea in summertime off the French coast at Brittany.

Scientists believe that the cause of this Palaeocene "supergreenhouse" was a massive release of carbon to the oceans and atmosphere, either from methane escaping from deepsea sediments or as organic carbon vaporized by violent volcanic eruptions that accompanied the opening of the North Atlantic Ocean. Methane produces CO_2 when it combines with oxygen. The cores show that the extra CO_2 in the atmosphere increased the greenhouse effect and warmed tropical temperatures by 4–5°C. The

ACE's findings show that Arctic temperatures also soared, rising from 18°C to 23°C. The work is a warning of how much the planet could heat up if we end up with a "runaway" greenhouse effect (see box below).

Meanwhile, sediments hauled up from the bottom of lakes in Mexico and Venezuela have revealed that a series of droughts struck Central America between 810 and 910AD, just before the great Mayan civilization collapsed. While climatic conditions are not widely accepted as the sole reason for the Mayans' downfall, the dates coincide with those when the cities were abandoned.

Tree rings

In parts of the world where there are strong seasonal changes in temperature or rainfall, trees form clear yearly growth rings. In spring, the extra moisture and warmer temperatures coax trees into a growth spurt that

---

RUNAWAY GREENHOUSE EFFECT

If a planet absorbs more energy from the sun than it radiates back into space, it experiences a "runaway greenhouse effect". In such circumstances, as the planet warms, a series of positive feedbacks cause it to warm more and the process eventually "runs away". Venus experienced such an effect three or four billion years ago. Over several hundred million years, the process turned it from a water-bearing planet to an inferno.

The runaway greenhouse effect works as follows. As sea surface temperatures rise above 27°C, evaporation releases a critical amount of water vapour into the atmosphere, one of the most efficient greenhouse gases. This absorbs infrared radiation from the sun, through the greenhouse effect, and warms the planet further. As the process repeats itself, the planet gets hotter and hotter.

On Earth, the only place that exhibits the signature of a runaway greenhouse effect is the Warm Pool of the Pacific. Here the sea surface temperature can be 30–32°C and the atmospheric humidity 10 kilometres above the Warm Pool can exceed 70 per cent. Nearby, where the sea surface temperature is cooler, the atmospheric relative humidity is much lower at 20 per cent. Scientists believe the drier neighbouring regions may stabilize the local runaway greenhouse effect.

Before human-induced climate change began warming the Earth, the planet's global average temperature was kept relatively constant by a balance of positive and negative feedbacks. It is possible that, as Earth warms as a result of human-induced climate change, positive feedbacks could begin to dominate, triggering a runaway greenhouse effect which would see the world's temperature spiral upwards.

The annual growth bands of trees yield information on past levels of solar radiation and hurricanes.

continues until the temperature falls in autumn. The size and density of cells within different parts of the annual growth band, and the width of each ring, vary according to changes to the weather within this growing season. By analysing variations in growth rings over time, dendroclimatologists can extract information about changes in climate that occurred throughout a tree's lifetime. In certain cases, they can match sections of wood from different trees found preserved in old buildings and rivers or lakes to build up chronologies dating back hundreds or thousands of years.

As well as looking at the appearance of tree rings, scientists use the concentrations of the isotope ^{14}C to monitor changes in the sun's energy. High-energy particles called cosmic rays constantly slam into the Earth's atmosphere and, as they do so, they create ^{14}C that trees take up as they grow. At times when the sun's energy output is high, there are fewer cosmic rays because charged particles from the star deflect them. This means tree rings produced at such times exhibit lower levels of ^{14}C.

Such a tree-ring analysis was used to demonstrate that the recent warming could not have been produced entirely by increased solar radiation. The 2004 study analysed fluctuating ^{14}C concentrations from trees dating back 11,400 years. It showed that the sun's energy output had been higher during the past 70 years than for the previous 8,000. However, this increased activity was not sufficient to account for the 0.5°C rise in global average temperature over the past 30 years. Solar radiation is linked to past changes in climate and society, however. The cooling associated with reduced solar output observed in trees around 476AD is blamed for the fall of the Roman Empire. The lower temperatures may have shortened the growing season in the northern hemisphere, causing crops to fail and prompting the Visigoths and other northern barbarians to head south for the more pleasant climes of the Roman Empire.

Scientists have also measured oxygen isotopes in tree rings to infer information about past hurricane activity. The Longleaf Pine (*Pinus palustris*) has shallow roots that suck in surface water when it rains. Hurricanes produce large amounts of water with a different ratio of ^{18}O to ^{16}O from that of dew or smaller storms. When the storm forms, the rain is enriched in the heavier isotope, but by the time it hits land, it contains more of the lighter isotope. By measuring different isotopes of oxygen present in the rings, Professors Claudia Mora and Henri Grissino-Mayer of the University of Tennessee identified periods when hurricanes hit areas of the south-eastern USA up to 500 years ago. As hurricane records date back only a century or so, the research could extend the historic hurricane databank, which would help researchers better understand hurricane frequency and intensity. In its 2007 Fourth Assessment Report (FAR), the IPCC concluded for the first time that hurricanes and cyclones are likely to become more intense, with higher peak wind speeds and heavier rainfall resulting from elevated sea-surface temperatures.

Corals

Corals build hard skeletons from calcium carbonate ($CaCO_3$) and some form annual rings, like trees, as they grow. In warm seas, corals grow faster, producing wide rings; in colder water they grow more slowly, giving rise to narrow bands. Both the width of the ring and the oxygen isotopes present are used by scientists to infer past water temperatures.

Scientists have also analysed collections of coral rubble dumped by storm surges to calculate when hurricanes took place in the past. Ridges in exposed areas tend to get destroyed by later storms, but those in

Coral reefs provide information on past ocean temperatures and have also yielded data on hurricanes.

sheltered locations on the leeward side of islands can build up one by one. Jon Nott, director of the Australasian Palaeohazards Research Unit at James Cook University, identified a series of 22 ridges dating back 5,000 years on Curacoa Island in the Great Barrier Reef, Australia. He deduced from the ridge locations and ages that ten times more category 4 or 5 "super-cyclones" had smashed into the Great Barrier Reef over this period than was previously believed. His findings suggested a category 5 cyclone should hit the Queensland coast about every 250 years. When Hurricane Larry battered Nott's home in Cairns in March 2006, he deduced from his calculations that the category 5 storm was overdue.

Ships' log books

Although tree rings and ice cores are useful for charting changes in climate on timescales of decades to millennia, researchers are particularly keen to examine past instances when the climate changed rapidly. To do so they need climate records spanning annual or even shorter periods, but reliable data about air movements over the oceans, which control much of our weather, barely existed before 1850. And by this time the industrial revolution was gathering speed and we were already set on a path towards CO_2 overload.

The Climatological Database for the World's Oceans (CLIWOC) project sought to solve this problem by analysing the logbooks of ships that crisscrossed the seas during the seventeenth century. From 1731 it was compulsory for British naval vessels to keep detailed records of the wind and weather they experienced, and by 1750 the practice was universal. The conditions were noted during each watch, roughly every four hours. It was an important job because few vessels at that time had the means to determine longitude accurately, and so relied on calculations of wind speed and force to estimate their pace and location.

The CLIWOC team, drawn from institutions in Europe and South America, trawled through more than 3,000 logbooks from some 5,000 voyages on the Indian and North and South Atlantic Oceans. They extracted more than 280,000 daily weather observations, taking care to correlate variations in language with modern-day Beaufort scale equivalents. As well as routine observations, the logs record events such as hurricanes, iceberg sightings, squalls and cloud cover. Because they are records of direct experiences with weather, rather than proxies, the database represents a century-long backward extension of oceanic and meteorological data being recorded today.

Canaletto paintings

The city of Venice risks being submerged by the combined effects of land subsidence and sea-level rise due to global warming. Regular tide monitoring started in 1872 and shows a relative sea-level rise of around 30 centimetres. However, in the past three decades the rise seems to have halted. Scientists wanted to know whether this was because subsidence had ceased owing to the regulation of underground water extraction or because of a temporary negative fluctuation in sea level. To find the

answer, Dario Camuffo and Giovanni Sturaro of the National Research Council at the Institute of Atmospheric Sciences and Climate, Padua, Italy, turned to an unlikely source of sea-level data: Canaletto's paintings.

The Venetian painter and his pupil Bernardo Bellotto made accurate representations of the Venetian cityscape using a camera obscura. This equipment operated like a modern-day film camera; the light beam travelled through a lens to be reflected by a mirror or prism on to a sheet of glass on which a paper or canvas was placed. The painter then filled in the details, producing a photographic painting. Many of the paintings show a line of brown-green *Laminaria* algae, the top of which is a good biological indicator of the high-tide level at the time. The paintings reveal that the high-tide mark has risen by some 60 centimetres since the first half of the eighteenth century.

The scientists used the tide data gathered before extensive water pumping began in 1930 to deduce a natural submersion rate of around 1.9 millimetres per year. This compared to a modern-day rate of 2.4 millimetres per year. They then used the paintings to consider fluctuations in submersion rates back three centuries. They found that the general trend was for a rising sea level, but that this was punctuated by several periods of stasis, including the present one. Today's situation reflected a two- to three-centimetre rebound of the aquifer, post-extraction, along with an increase in atmospheric pressure over the past four decades, equivalent to a one-centimetre decrease in sea level. The scientists concluded that the current halt in the sea-level rise could well be a temporary fluctuation and suggested that a likely trend was for a rise in sea level somewhere between the 1.9 millimetres per year past natural rate and 2.3 millimetres per year past overall rate, as deduced from the paintings. Venice has lost 60

Lines of *Laminaria* algae around the base of buildings in Canaletto's paintings have helped scientists deduce information on past sea levels.

centimetres around the base of its historical buildings to the sea and the walls are at risk of rapid decay. Therefore, the scientists suggested safeguarding measures to be vital if Venice were to be preserved for future generations.

The advance of wild vines

Wild vines grew in Britain as far back as 50 million years ago and grapes were first domesticated here some 5,000 years ago. By mapping the ebb and flow of British vineyards over the past two millennia, Professor Richard Selley of Imperial College London has demonstrated that the changing distribution of vines can be a useful indicator of fluctuating temperatures. In his book *The Winelands of Britain, Past Present and Prospective*, he shows that vineyards flourished during Roman and Medieval times south-east of a line linking the Humber and Severn estuaries. However, during the Little Ice Age of 1300 to 1850, they were restricted to the south-east corner of England. In the present post-industrial revolution warm phase, vines have almost reclaimed their former lands.

OUR FUTURE WORLD

Accurately forecasting how future changes to our climate will affect us depends on all kinds of things, from how quickly the world's population swells to changes in gross domestic product (GDP) and the speed at which we are able to introduce new technologies to cut our use of fossil fuels or suck the greenhouse gases from the atmosphere. In 2000, the IPCC published a Special Report on Emissions Scenarios (SRES) containing predictions on the various shapes of our future world. The aim was to make it easier for people to assess the consequences of potential future greenhouse gas concentrations and create feasible strategies for mitigating, or adapting to, the coming changes. The scenarios consider the period from 1990 to 2100 and were used in the Fourth Assessment Report (FAR) to express likely climatic outcomes. None of the scenarios explicitly assumes that the United Nations Framework Convention on Climate Change or the emissions targets of the Kyoto Protocol (see Chapter 4) will be implemented.

There are around 40 scenarios clustered under the following four broad groups:

• **A1: Fast economic growth, new and efficient technologies**
This future world is defined by very rapid economic growth, a global population that peaks in the mid-century then declines, and the rapid introduction of new and more efficient technologies. Three sub-groups are defined according to their technological emphasis. A1F1 is fossil intensive, A1T is non-fossil-fuel reliant and A1B represents a balance of

Geology and climate both bear an influence on where vines grow well. By analysing the underlying rocks across Britain and assessing the likely changes that global warming will bring to our climate, Professor Selley identified four potential regions where vines will thrive in the future. These are: wine lands abandoned during the Little Ice Age on the greensand hills of Surrey; reborn Roman and medieval wine lands on chalk downland and river terrace gravels along the Thames; new wine lands developed in the Weald and among the rocky rivieras of south-west England and Wales; and future wine lands, if global warming continues, on the southern slopes of the Lake and Peak Districts, followed by the Southern Uplands, Grampians and Great Glen of Scotland.

Predicting the future

Computer modellers use the data on past climates gathered from proxy sources, combined with contemporary records of variables such as temperature, humidity and greenhouse gas levels, to test that a particular

energy sources. The A1B and A1F1 scenarios are the ones that presently seem to most closely match changes happening in the real world.

• **A2: Slower technological growth, rising population**
The A2 world is a very heterogeneous one. The underlying theme is self-reliance and preservation of local identities. Fertility patterns across regions converge slowly, resulting in a continuously increasing population. Economic development is mostly regionally oriented, while per capita economic growth and technological changes are slower than in other storylines.

• **B1: Service economy, global solutions to a sustainable world**
The B1 clutch of scenarios describes a convergent world with the same global population trend as in the A1 storyline, but with rapid change in economic structures towards a service and information economy, with reductions in material intensity and the introduction of clean and resource-efficient technologies. The emphasis is on global solutions to economic, social and environmental sustainability, including improved equity, but without additional climate initiatives.

• **B2: Local solutions to a sustainable world**
These scenarios describe a world in which the emphasis is on finding local solutions to economic, social and environmental sustainability. This world has an increasing global population (but one that is expanding at a rate lower than A2), intermediate levels of economic development, and less rapid but more diverse technological changes than in the B1 and A1 storylines. This outlook is also orientated towards environmental protection and social equity, but focuses on local and regional levels.

model can faithfully reproduce past conditions. Then, starting with today's climate, they crank up concentrations of CO_2 or methane to see what such atmospheric inputs might have in store for humanity.

Models are essentially simplified mathematical representations of the Earth's climate system. Early models, developed in the 1970s, were able to create only basic representations of the planet. However, thanks to our increased knowledge about the real world, coupled with greater computing speed and power, today's state-of-the-art models are much more complex. They reproduce the major components of the climate system – the atmosphere, oceans, land surface, ice fields and biomass – along with processes that go on within and between them. They enable scientists to investigate processes such as how clouds affect the heating and cooling of the planet, the varying times taken for the land and oceans to heat up, and how the intricate workings of Earth's chemistry and carbon cycles control how much CO_2 exists in the atmosphere. The models range from powerful number-crunching General Circulation Models that aim to predict variations to climate across the whole globe, to simple models investigating likely changes to a single aspect of the world's intricate weather machine.

The latest model in operation at the Met Office Hadley Centre in the UK, HadGEM1, is typical of current state-of-the-art models. It divides the world into 135-kilometre by 135-kilometre boxes, with 38 vertical layers representing the atmosphere and 40 levels describing the ocean. As some processes, such as those related to clouds or ocean convection, take place at much smaller scales, their average effects are approximated using a technique called parameterization. This involves assigning fixed numbers to processes, when their value is not known for certain. For example, based on knowledge of the temperature and humidity in a particular cube, the model-makers might estimate how much cloud and rain there is in each box. The IPCC's Fourth Assessment Report (FAR) drew heavily on model simulations to make its predictions for how the climate is likely to change in the future. Even since the panel made its third assessment in 2001, a large number of new simulations from a broader range of models have become available; the latest assessment is based on 19 models, almost double the previous number.

The world's rising fever

The IPCC's FAR assessment confirmed that recent warming of the climate system is unequivocal, with evidence coming from records of global air and ocean temperatures, the widespread melting of snow and ice and rising sea levels. The IPCC considers the total temperature increase from 1850–1899 to 2001–2005 is 0.76°C, with the rate of warming over the past 50 years nearly double that for the last 100 years. Observations made since 1961 show that the average temperature of the global ocean has increased to depths of at least 3,000 metres. The amount

COMPUTER USERS UNITE TO HELP SCIENTISTS PREDICT THE FUTURE

People around the world have been helping climate scientists predict the Earth's weather in the twenty-first century by loaning them the number-crunching power of their computers. Climate models predict significant changes to the Earth's climate in the coming century but there is a huge range in what they forewarn. The climateprediction.net project wanted to run hundreds of thousands of state-of-the-art climate simulations at locations across the world, with slightly different physics to represent the whole range of uncertainties for all possible permutations of climatic variables.

Some 259,700 participants from 171 countries agreed to let the project use their PCs when they were not using them. All the participants had to do to take part was download a version of the Met Office's state-of-the-art climate model that would run on their computer when the processor was not in use for other jobs.

The project was made up of three separate experiments. One was designed to explore the model being used, the second to see how well the models replicated past climates, and the third to produce a forecast for the twenty-first century. Every model distributed contributed to all three experiments. Each simulation differed from all the others in three ways: the initial conditions it started from; the attributes which forced it to be in one particular climate state (these are processes, such as volcanic eruptions, that you might not consider to be part of the climate but which can have a large influence on weather); and the parameters, or approximations, that made up the actual model.

The first results of the experiment became available in early 2007. They suggest that the UK could be warmer by as much as 4°C by 2080, agreeing with other models. Areas at high latitudes such as Alaska and north-west Canada are likely to fare the worst in terms of temperature rise. This is because the loss of white ice from these areas will cause less of the sun's radiation to be reflected back into space, so increasing warming. The Sahara and Central Asia are also expected to warm up considerably, because their landlocked locations mean they are not subjected to the temperature-ameliorating effects of the oceans. The experiment predicted that New Zealand would warm much less than landlocked areas as the nation is surrounded by water, and water takes much longer to warm up than land.

Climateprediction.net runs on a software platform called the Berkeley Open Infrastructure for Network Computing (BOINC). You can read more about the experiment at: http://www.bbc.co.uk/sn/climateexperiment/

THE EVOLUTION OF CLIMATE MODELS

1950s: The development of computers leads early meteorologists to seek new ways to predict the weather.

1955: Norman Phillips, in Princeton, USA, develops the first true general circulation model (GCM) on a computer. The atmosphere is essentially represented as a cyclinder, with figures changed to represent the addition of heat in the lower half. The resulting calculations produce a plausible jet stream.

1964: Yale Mintz and Akio Arakawa, at the department of Meteorology at the University of California, Los Angeles, produce a climate computed for an entire globe. Its atmosphere is represented by just two layers, but it includes realistic geography such as mountains, oceans and ice cover.

1965: Joseph Smagorinsky and Syukuro 'Suki' Manabe, at the Geophysical Fluid Dynamics Laboratory near Washington, produce the first reasonably complete 3D model of the Earth's climate system. It reproduces the stratosphere, a zone of rising air at the equator representing the doldrums, and a subtropical band of deserts. However, it is highly simplified and contains no geographical features.

1967: Using a single one-dimensional column to represent the make-up of the atmosphere averaged over the entire globe, Manabe's team conducts an experiment to see what would happen if CO_2 in the atmosphere doubled. The computer predicts the temperature would rise by roughly 2°C.

Early 1970s: Weather disasters and the energy crisis put greenhouse warming on the political agenda. The controversy of whether computer models are correct in predicting how CO_2 emissions might influence temperature enters public debate.

1979: A panel appointed by the National Academy of Sciences meet at Woods Hole, USA, to compare predictions made by two separate GCMs, built using different physical approaches and computational methods. One is the creation of Manabe and colleagues and the other the brainchild of Jim Hansen's team at NASA's Goddard Institute for Space Studies in New York City. The panel finds the models agree that the world will get warmer if CO_2 levels rise. Hansen's GCM predicts a 4°C rise for doubled CO_2; Manabe's forecasts a 2°C rise. The panel concludes that they have rather high confidence that as CO_2 reaches this level the planet will warm up by about 3°C plus or minus 50 per cent. They report that: "We have tried but have been unable to find any overlooked or underestimated physical effects" that could reduce the warming.

Early 1980s: Several groups develop more realistic models. These are reasonable representations of the Earth's geography, including oceans capable of exchanging heat with the atmosphere. They are known as "coupled ocean-atmosphere" models. The models agree with earlier, simpler ones, that increased CO_2 will raise the global temperature.

1983: The US National Center for Atmospheric Research (NCAR) develops the first official Community Climate Model and publishes its computer source codes, so that it can be refined by scientists working at different institutions and studying a range of disciplines.

1985–88: Kirk Bryan and a collaborator try out a coupled ocean-atmosphere model with its CO_2 levels set at four times higher than the present-day concentration. They find signs that the world's oceanic conveyor belt, the Meridional Overturning Circulation, could switch off. Three years later, Manabe runs a simulation which shows that, even at present CO_2 levels, it might be possible for the ocean-atmosphere system to settle into one of two states; one with the conveyor on and one with it off. Some experts begin to worry that global warming might shut down the Earth's oceanic central heating system.

1989: A group of scientists from the USA, Canada, England, France, Germany, China and Japan feed the same external forcing into 14 different models using sea-surface temperature to represent climate change. The predictions agree well for clear skies but fall down badly when clouds are introduced.

Late 1980s: Instruments on board satellites begin generating a flow of accurate, contemporary data on incoming and outgoing radiation, cloud cover and other climatic parameters. The data shows that clouds have an overall cooling effect on the planet, which highlights the importance of including cloud data in climate models. Scientists use the new global data bank as a backdrop against which to set predictions of future change. The modelling of aerosols improves.

1991–95: Jim Hansen's team uses the eruption of Mount Pinatubo in the Philippines, which generates a sulphuric acid haze in the stratosphere around the world, as a means to test the accuracy of his group's model. They forecast a global cooling lasting a couple of years, which four years later proves to be accurate.

1998: Several different groups, using a plethora of coupled atmosphere-ocean models, produce roughly similar simulations of the ice age climate, proving that the models can reproduce climates other than the present one.

Late 1990s: Some specially designed regional models are now able to produce reasonably accurate representations of the El Niño/Southern Oscillation. Now armed with more superior computing power than in the 1980s, Manabe's group shows that a steady increase in CO_2 could seriously weaken the Meridional Overturning Circulation within a few centuries.

2001: Two groups use coupled models to match the rise in temperature detected in the world's oceans. Meanwhile, the IPCC's TAR relies heavily on the results of modelling to predict future outcomes of climate change, lending the community increasing credibility.

2004: After two decades of refinements, NCAR's Third Community Climate System Model is released, its new name reflecting the increasing complexity of models.

2005: Computer modellers are now sufficiently confident to declare that temperature measurements over the previous four decades give an unequivocal signature of human-induced global warming. The pattern of warming in different ocean basins matches what models predicted the rise would be, after some delay, from solar energy trapped by greenhouse gases emitted by industry.

2006: NCAR's continually developing Community Climate System Model has by now undertaken 12,000 simulations since its conception. The Holy Grail for modellers is to provide a fully integrated Earth system model that can be simulated every 15 minutes for centuries. Models around the world continue to be honed towards this goal; Oak Ridge National Laboratory announces a five-year $1.2 million project with NASA's Goddard Space Flight Center involving a satellite that will allow it to track down carbon dioxide emissions to 75-kilometre grids around the world. Meanwhile, a new semi-empirical method used to predict future sea-level rise suggests current estimates produced by models and used in the 2001 IPCC report may be half what they should be.

2007: The IPCC's FAR notes that, since the panel made its third assessment in 2001, a large number of simulations from a broader range of models have become available. It uses these, together with information from observations, to produce estimates for aspects of future climate change. It concludes that temperatures are most likely to increase by 1.8–4°C by the end of the century and that sea levels will probably rise by 28–43 centimetres, but could climb by 59 centimetres. Months later, scientists publish results of a study combining the approaches of weather forecasters, who typically look a few days ahead, with climate modellers, who make projections to the end of the century. Their resulting model, which predicts changes to 2015, fills a long-standing gap in climate prediction techniques.

of water vapour in the atmosphere has also increased since at least the 1980s; this would be expected, as water evaporates more readily under warm conditions.

Information revealed by ice cores and other climate proxies suggests the average northern hemisphere temperatures during the latter half of the twentieth century were very likely (which in IPCC-speak means there is a greater than 90 per cent chance) higher than during any other 50-year period in the past 500 years and likely (a more than 66 per cent chance) to have been the highest in the past 1,300 years. Since the IPCC's first report in 1990, predictions for temperature increases between 1990 and 2005 were in the range 0.15–0.3°C per decade. The recorded increase of 0.2°C per decade shows that these forecasts were pretty spot on, lending credibility to the latest predictions. The forecasts for the magnitude of warming expected by the end of the century span the range from 1.1°C to 6.4°C, relative to 1980–1999 temperatures. The best estimate for the B1 scenario is for a rise of 1.8°C, while inhabitants of the A1B world can expect a 2.8°C increase. The A1F1 future could bring about a massive jump in temperature of 4.0°C.

A rise in temperature at the higher end of this scale is likely to have a crushing impact on the world's climates, and affect industries from agriculture to tourism. Because greenhouse gases linger in the atmosphere for considerable lengths of time, any action we take now will not have that much impact in the next 40 to 50 years. Even if the concentrations of all greenhouse gases and aerosols had been kept to a constant at year 2000 levels, we could still expect a further warming of about 0.1°C per decade. Given that greenhouse gases are presently haemorrhaging into the atmosphere at a quickening pace, doing nothing in the next 10 or 20 years could well condemn the planet to temperature rises of 4°C or more in the next century, causing unprecedented changes to the planet's natural systems (see box: Climate impacts at 2°C, 3°C and 4°C, page 125–127).

One recent project analysed the latest climate change predictions generated by 20 global climate models using the A1B, A2 and B1 emissions scenarios. The researchers produced a Regional Climate Change Index, based on forecast changes in rainfall and temperature, to identify regions likely to have the greatest response to climate change. The scientists, from Abdus Salam International Centre for Theoretical Physics in Trieste, Italy, identified potential hotspots to be: the Mediterranean and north-eastern Europe, high-latitude areas in the northern hemisphere, Central America, southern equatorial Africa, the Sahara and eastern North America. According to their results, the Mediterranean and northern Europe can expect less rainfall overall, but more variability, making droughts and flash floods more likely. North-eastern Europe is set to have much more snow, while eastern North America can expect irregular weather patterns.

The results of a similar project, carried out by scientists at the Swiss Federal Institute of Technology in Zurich, were published in early 2007. The researchers divided the world into 375-kilometre by 375-kilometre squares

and, for each of the nine climate-change indicators, identified extreme events that had taken place between 1961 and 1990 that would normally have been expected to occur once in 20 years. Using three different climate models, the scientists then computed the likely changes in the frequency of extreme events, given a mid-range forecast for greenhouse gas emissions, for the period 2071 to 2100.

A challenge for climate change scientists is how to predict the influence melting ice, such as that in Antarctica, will have on sea-level rise.

They weighted each change to give a number between 0 and 19 for each square, with 0 representing a situation in which all nine climate indicators remain as one-in-20-year events and 19 representing all nine indicators becoming annual events. Plotting the results for different indicators revealed some startling potential changes. The temperature map showed that one-in-20-year elevated temperatures are set to become the norm across much of the world by the end of the century. Meanwhile, Antarctica and the Arctic are 13 times more likely to experience wet years, while the Amazon rainforest and Congo Basin are 13 times more likely to suffer droughts. Europe, the USA and Australia are likely to fare better, but are still on course to experience more extreme events.

A swelling tide of evidence

One of the great challenges for climate scientists is to predict how much the sea level will rise around the world in response to global warming. The fate of coastal cities, including New York, Tokyo and Shanghai, numerous islands, and low-lying countries such as Bangladesh and the Netherlands, hangs on getting these predictions right, so that policymakers can develop strategies to protect valuable land and people's homes. The ocean has been absorbing more than 80 per cent of the heat added to the climate system in recent years. Such warming causes water to expand, contributing to sea-level rise. Other elements with the potential to contribute to higher oceans are the melting of glaciers and the vast ice sheets of Antarctica and Greenland.

The world's oceans are already rising. Global average sea level rose 1.8 millimetres per year between 1961 and 2003, with the rise speeding up to 3.1 millimetres per year between 1993 and 2003. This brings the total sea-level rise throughout the twentieth century to around 17 centimetres. The IPCC's best estimates for the likely sea-level rise by the end of the century range from 18 centimetres to 59 centimetres. However, the models used to generate these figures do not include the possible effects of changes in the rate of melting of the Greenland and Antarctic ice sheets. This is because we do not yet know enough about the physical processes controlling the way ice sheets melt to include them in models. Some scientists are concerned that this omission means the latest predictions are too conservative.

Researchers at the Potsdam Institute for Climate Impact Research in Germany took the predictions for CO_2 emissions, temperature and sea-level rise given in the IPCC's Third Assessment Report (TAR) in 2001 and compared them to the changes actually observed between 1990, the most recent year for which data was available at that time, and the present time. They found that the changes in CO_2 closely mirrored the predictions made in the 2001 report and that temperature also matched the forecast, but was close to the top of the predicted range. However, the "best

Already prone to flooding, Bangladesh is highly vulnerable to sea-level rise.

estimate" prediction for sea-level rise, of less than 2 millimetres per year, was lower than the observed rise of 3.3 millimetres per year between 1993 and 2006.

Evidence from palaeoclimatic records shows that the last time polar regions were significantly warmer than at present was 125,000 years ago, at the peak of the last interglacial. At this time, polar temperatures were between 2°C and 3°C warmer than at present. Melting of the polar ice led to sea-level rise in the region of three to six metres. One record that demonstrates this rise is a fossil coral reef in Foul Bay, near Margaret River in western Australia. The reef dates back to between 125,000 and 128,000 years ago. Its position at around 2.5 metres above the present tide level means that it must have been formed when the sea was at least three to four metres higher than its present level, and possibly six metres higher.

Scientists know that the expansion of the oceans through thermal warming can account for only around 50 centimetres of sea-level rise, so the melting of the ice sheets must have contributed to the ocean's elevated level. A sea-level rise of four metres could be explained as melting purely from the Greenland ice sheet, Iceland and other Arctic ice fields plus small contributions from northern hemisphere glaciers. However, a rise of six metres, would require substantial melting of Antarctica's ice in addition to the Arctic input. If, as some experts suggest, polar warming by 2100 reaches levels similar to the peak of the last interglacial, this would induce much greater rises in sea level than forecast in the IPCC's 2007 FAR.

Farming in a warmer world

How individual crops respond to a changing climate depends on the species, cultivar, soil conditions, reactions to higher levels of CO_2 and the ways in which farmers opt to manage their land. A warming world has the potential to both increase and decrease crop yields. A few degrees of warming will probably boost yields at temperate latitudes, but heat greater than this will generally cut yields. The FAR predicts that moderate climate change in the early decades of the century will increase yields of rain-fed agriculture in parts of North America by between five and 20 per cent. Actions such as altering sowing times and choosing those cultivars that thrive in the heat could help to avoid dwindling harvests. In the tropics, the picture is less positive. Here, dryland agriculture predominates, and the temperature is already close to the maximum that some crops can tolerate. Yields are likely to decrease with even a minimal warming, and increasingly sparse rainfall will have an even more adverse affect. Changes to management methods will probably make little difference at low latitudes. More frequent droughts and floods will compound the negative impact of temperature on crop yields. Most studies suggest that a mean annual temperature increase of 2.5°C or greater will push food prices up.

The impacts of higher sea levels around the globe

Thirteen of the 15 largest cities are located in coastal areas and the trend is for more people to move to urban areas. Experts estimate that by the year 2030, 50 per cent of the world's population will live within 100 kilometres of the coast.

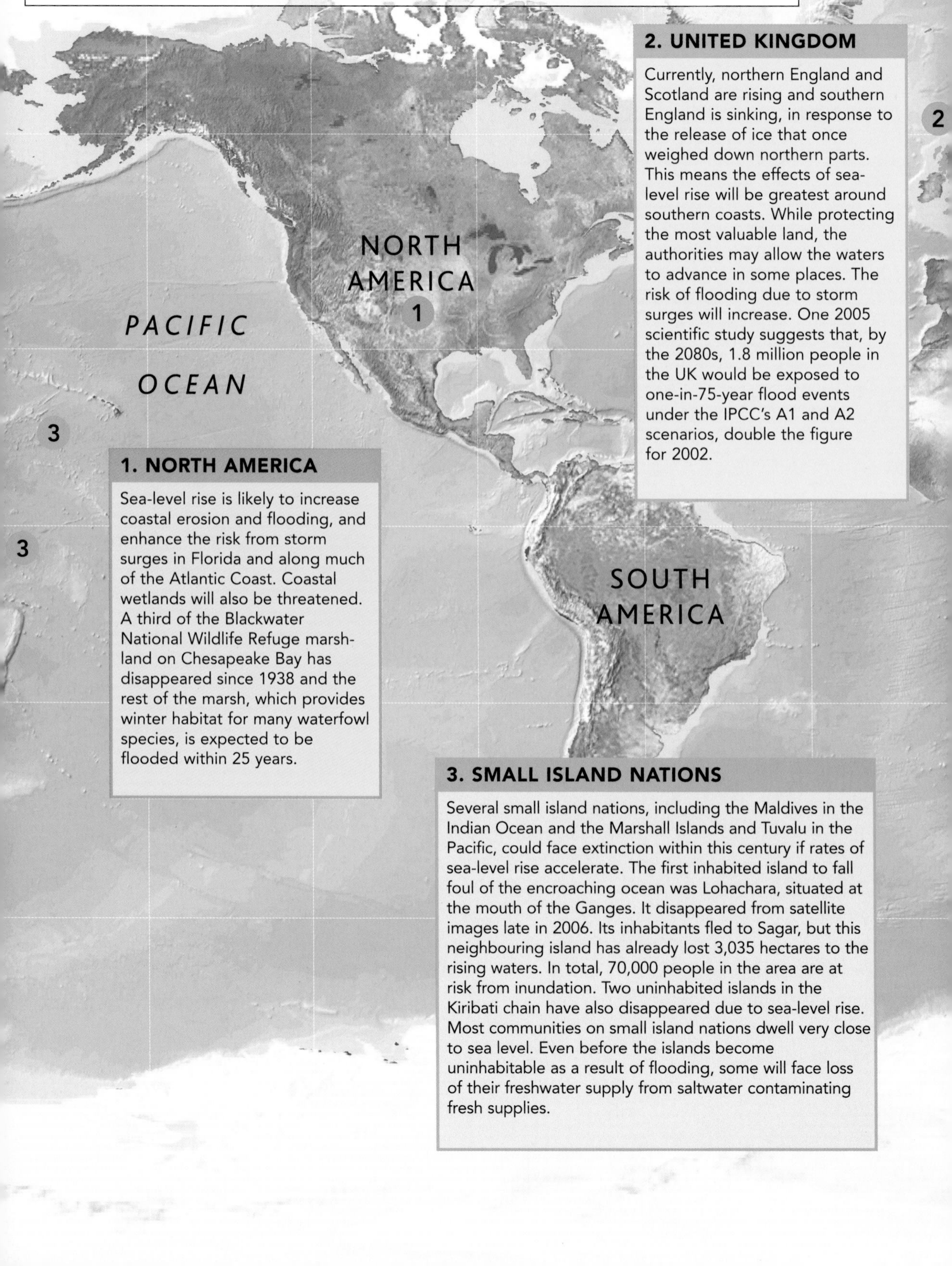

2. UNITED KINGDOM

Currently, northern England and Scotland are rising and southern England is sinking, in response to the release of ice that once weighed down northern parts. This means the effects of sea-level rise will be greatest around southern coasts. While protecting the most valuable land, the authorities may allow the waters to advance in some places. The risk of flooding due to storm surges will increase. One 2005 scientific study suggests that, by the 2080s, 1.8 million people in the UK would be exposed to one-in-75-year flood events under the IPCC's A1 and A2 scenarios, double the figure for 2002.

NORTH
AMERICA
1

PACIFIC

OCEAN

3

3

2

1. NORTH AMERICA

Sea-level rise is likely to increase coastal erosion and flooding, and enhance the risk from storm surges in Florida and along much of the Atlantic Coast. Coastal wetlands will also be threatened. A third of the Blackwater National Wildlife Refuge marsh-land on Chesapeake Bay has disappeared since 1938 and the rest of the marsh, which provides winter habitat for many waterfowl species, is expected to be flooded within 25 years.

SOUTH
AMERICA

3. SMALL ISLAND NATIONS

Several small island nations, including the Maldives in the Indian Ocean and the Marshall Islands and Tuvalu in the Pacific, could face extinction within this century if rates of sea-level rise accelerate. The first inhabited island to fall foul of the encroaching ocean was Lohachara, situated at the mouth of the Ganges. It disappeared from satellite images late in 2006. Its inhabitants fled to Sagar, but this neighbouring island has already lost 3,035 hectares to the rising waters. In total, 70,000 people in the area are at risk from inundation. Two uninhabited islands in the Kiribati chain have also disappeared due to sea-level rise. Most communities on small island nations dwell very close to sea level. Even before the islands become uninhabitable as a result of flooding, some will face loss of their freshwater supply from saltwater contaminating fresh supplies.

OCEAN

4. EUROPE

The greatest potential losses from sea-level rise are in the east of England, the Po Delta of northern Italy and along a swathe of coast running from Belgium, through the Netherlands and north-west Germany into western and northern Denmark. Some seaside communities in the Netherlands, England, Denmark, Germany, Italy and Poland are already below normal high-tide levels, and wider areas are vulnerable to flooding from storm surges. Such threats may well increase in the future. Regional and local sea-level rise in Europe will differ from the global average because of vertical land movements caused by the release of pressure from melted ice following the last glaciation, shifts of geological plates, and subsidence.

5. ASIA

Sea-level rise and an increase in the intensity of tropical cyclones could displace tens of millions of people in low-lying coastal areas of temperate and tropical Asia. It would also threaten mangroves and coral reefs.

6. VIETNAM

Vietnam's Mekong River delta, which plays an important role in the Vietnamese economy, has been severely affected during this century by unusually large floods. Further sea-level rise is likely to increase the risk of flooding and extend the area affected by saltwater further inland.

ASIA

AFRICA

4 EUROPE

INDIAN OCEAN

OCEANIA

9. SINGAPORE

As a small, low-lying nation, Singapore is highly vulnerable to sea-level rise. A study into the varied costs of protecting it or allowing the sea to inundate areas found that strengthening defences to protect vulnerable coastal areas was likely to be the cheapest option. It calculated that the annual cost of protecting Singapore would rise over time and range from US$0.3–5.7million by 2050 to US$0.9–16.8million by 2100.

7. AFRICA

Frequent inundations and increased erosion from sea-level rise threaten coastal settlements including those in the Gulf of Guinea, Senegal, Gambia, Egypt, and along eastern parts of the southern African coast. A sea-level rise of 50 centimetres would displace some 1.5 million people in Alexandria and Port Said in Egypt – a figure that includes half the population of Alexandria.

8. BANGLADESH

Bangladesh, one of the world's poorest and most crowded nations, is highly vulnerable to sea-level rise. Storm surges already regularly make people's lives a misery. Catastrophic floods in the past have resulted in damage as far as 100 kilometres inland. A sea-level rise of 45 centimetres could result in 15,668 square kilometres of land being lost, over ten per cent of the country. This would displace 5.5 million people.

ANTARCTICA

Away from the parched land, global climate change will affect the temperature and sea level of the oceans, melt sea-ice, and change salinity, acidity, wave patterns and the circulation of ocean currents. Such changes will have an impact on the health and distribution of marine plants and animals. Minute single-celled plants called phytoplankton form the foundation of the ocean food chain. Areas rich in phytoplankton attract small fish, which in turn encourage larger predators to gather. These areas are dynamic, shifting with passing storms and ocean current patterns. Areas where reliable, upwelling currents bring such nutrient-rich waters and their feeding shoals close to shore provide rich pickings for fishermen. Changes to natural circulation patterns, which whisk plankton and fish away from the coast, could have a devastating impact on local fisheries.

Climatic changes may shift nutrient-rich ocean currents, reducing fish stocks in some areas.

The geographic shift of traditional species in line with temperature changes will also take their toll. Cold- and cool-water species will see their habitats reduced, while the range of warm-water fish will expand. Generally, there will be a trend for species to shift poleward, as waters become warmer at higher latitudes. The expansion of marine aquaculture could partially compensate for reductions in catches of ocean fish. Marine aquaculture production has more than doubled since 1990; in 1997 it represented around 30 per cent of fish and shellfish produced for our dining tables. However, increased temperatures may reduce stocks of fish such as herring and anchovies, which are used to feed cultured stocks. The abnormal warmth could also cut the amount of oxygen dissolved in the seawater. This is known to encourage disease to spread among farmed and wild stocks.

Africa

Climate change will likely worsen food security in Africa. Per-capita water availability has diminished by 75 per cent on the continent over the last 50 years, mainly due to the swelling population. The FAR predicts that by 2020, between 75 million and 250 million people will face increased water stress due to climate change. The Horn of Africa and Southern African Development Community (SADC) regions are likely to suffer recurrent droughts. Variations in climate will bring about pronounced fluctuations in water levels in inland water bodies such as the East African Lakes, causing hardship for communities that rely on them to provide valuable protein. Offshore, the coastal marine fisheries of Namibia, Angola and South Africa may be affected by changes to the Benguela current. This ecosystem is the strongest wind-driven coastal upwelling system known. It supports large numbers of fish, crustaceans, seabirds and marine mammals and generates thick shoals of herrings, sardines and anchovies. Climate change could make the nutrient-rich current more variable and unstable, affecting commercial fishing.

Asia

Asia holds more than 60 per cent of the world's population, and climate change will heap more stress on its already stretched resources. Food insecurity is a major concern. Although crop yields could rise by 20 per cent in East and South-east Asia, they may decrease by as much as 30 per cent in Central and South Asia. China's agricultural output is likely to shrink, while water shortages and high temperatures could adversely affect wheat and rice harvests in India. Warmer and wetter weather may make diseases such as wheat scab and rice blast more prevalent in temperate and tropical regions. Farmers may be able to offset some of the negative impacts by changing planting seasons to take advantage of rainy spells and avoiding destructive events such as typhoons. However, according to the FAR, the availability of fresh water in Central, South, East and South-east Asia will decrease with climate change, particularly in large river basins. Surface run-off, where water from rain, melting snow or irrigation washes over soil into rivers rather than percolating down into the water table, is predicted to decrease drastically in arid and semi-arid regions. A 2°C increase in temperature coupled with five or ten per cent less rainfall will greatly reduce surface runoff in Kazakhstan and have serious implications for agriculture and livestock farming. Given Asia's swelling population and increased demand for higher living standards, the reduced availability of water could affect more than a billion people by the 2050s. Asia produces 80 per cent of all farmed fish, shrimp and shellfish. Marine life here is already vulnerable to shifts in plankton blooms prompted by changes in temperature brought about by the El Niño/Southern Oscillation (ENSO). More frequent ENSO conditions, along with increased storm surges, could reduce the amounts of fish netted.

Australia and New Zealand

More frequent droughts will slash agricultural productivity and increase the risk of forest fires in parts of southern and eastern Australia and eastern New Zealand. Parts of western and southern New Zealand may benefit initially owing to a longer growing season, less frost and increased rain.

IS WATER THE NEW OIL?

Viewed from space, the Earth is a blue, watery planet. It looks as if we have a never-ending supply of this life-giving liquid, but only three per cent of it is not salty. Two-thirds of this fresh-water is locked up in ice caps and glaciers. Of the remaining one per cent, one-fifth is in remote, inaccessible areas and a good deal of the rest arrives when it is least wanted, as monsoonal deluges and floods. The result is that humans are able to exploit only 0.08 per cent of all the world's water. With the global population predicted to rise to 8.9 billion by 2050, and as climate change brings droughts to new areas, the battle for dwindling supplies is likely to stimulate new regional tensions and wars. In 1999, the United Nations Environment Programme (UNEP) reported that 200 scientists in 50 countries had identified water shortage and global warming as the two greatest challenges for the world in the new millennium.

Seventy per cent of our water is used in agriculture, and much of this is wasted through inefficient irrigation schemes. Our dietary tastes also make us wasteful of water. It takes 2,000 to 5,000 litres of water to produce a kilogram of rice and 11,000 to produce a single hamburger. Global water consumption rose six-fold between 1990 and 1995 and, with ever more people seeking Western lifestyles and diets, that figure continues to rise. The result has been an increase in the number of countries suffering from either water stress (defined as there being less than 1,700 cubic metres of water per person per year) or water scarcity (less than 1,000 cubic metres per person per year). Approximately 1.7 billion people, around one-third of the world's population, presently live in countries that are water-stressed. In 1955, there were seven countries considered to be suffering from water scarcity; by 1990 a further 13 had been added. The UN anticipates that a further 14 countries will slip from water stress to scarcity by 2025. The World Water Council has predicted that by 2020 a further 17 per cent more water than is available will be required to feed the world.

Of the countries likely to be considered water-scarce by 2020, just over two-thirds are located in the Middle East. Here, five per cent of the world's population rely on just one per cent of its water. Figures of water withdrawal as a percentage of renewable water supplies are among the highest in the world, while renewal rates are low because of the region's arid climate. There are signs that water could take over from oil as the main source of conflict in the region. In the 1960s, Israel, Syria and Jordan took part in cross-border raids on machinery being used for water schemes, activities that culminated in the Six-Day War. And the first operation undertaken by Yasser Arafat's Fateh organization in 1964–65 was against the national water carrier in Israel. More recently, Turkey almost came to blows with Syria in 1998 over the former's plans to build dams on the River Euphrates. As climate change and rising populations conspire to increase the pressure on water supplies in the coming years, water wars are likely to affect not only Middle Eastern countries, but Africa, India, China and Bangladesh.

The possibility of global warming pushing water-stressed countries towards conflict has not gone unnoticed by governments. In early 2006 Britain's Defence Secretary, John Reid, warned of the geopolitical and social consequences that global warming could have on a world whose population was rising at the same time as its water resources were dwindling. He signalled that Britain's armed forces would need to be prepared for humanitarian disaster relief, peace-keeping and warfare to deal with dramatic social and political consequences of climate change. Some of the least equipped countries are the ones most likely to be forced to deal with flooding, water shortages and agricultural land turning to desert.

However, pests and diseases may spread more readily. The region's reliance on agricultural and timber exports make it sensitive to climate-induced changes in production elsewhere in the world. Changes to winds and currents that influence the location of nutrient-rich upwellings could affect fisheries.

Europe

Nearly all regions of Europe will be negatively affected by some of the impacts of climate change; southern and Arctic Europe are particularly vulnerable. In northern Europe, the effects of change on agriculture are likely to be mostly positive, with gains to crop yields. Any positive impact experienced in southern countries will probably be offset by the risk of water shortages and the shortening of growth periods in grain crops caused by higher temperatures. Poleward shifts of some marine species could affect fisheries already stressed by overexploitation.

Central and South America

The ENSO cycle has a big influence on climate here. El Niño ushers in dry conditions to north-east Brazil, northern Amazonia, the Peruvian/Bolivian Altiplano and the Pacific coast of Central America. Recent El Niño events have brought drought to Mexico and rain to southern Brazil and north-western Peru. La Niña years are associated with heavy rain and flooding in Colombia and drought in southern Brazil. If El Niño or La Niña conditions increase with global warming, these weather patterns will become more prevalent. In drier areas, climate change is likely to prompt salinisation and desertification of agricultural land. Studies in Argentina, Brazil, Chile, Mexico and Uruguay based on general circulation models and crop simulations predict that yields of maize, wheat, barley and grapes will drop. This could have serious economic consequences given that agriculture contributes 10 per cent of the GDP of Latin America and employs some 40 per cent of the economically active population. However, temperate regions may benefit from higher soybean yields.

North America

Climate change will deal North America pluses and minuses. In the Midwest and Canadian Prairies, there is an increasing chance of severe drought and the potential for large declines in the water flowing in the region's rivers. These influences will make irrigation essential in some areas previously blessed with reliable rainfall. Canada may have opportunities to expand agricultural production northwards. At sea, climate variations could affect the productivity of several North American fisheries in the Pacific, North Atlantic, Bering Sea and Gulf of Mexico regions.

Small island nations

Such countries have a limited amount of freshwater and arable land. The FAR predicts that, by the middle of this century, the water resources of many small islands, including some in the Pacific and Caribbean, will become insufficient to meet demand during periods of low rainfall. In many places, farming is concentrated in low-lying areas near coasts that are already vulnerable to inundation by the sea. Saltier soil will affect the success of staple crops such as taro. Although often small scale, fishing tends to make an important contribution to the diet of islanders. Many breeding grounds and habitats for fish and shellfish, such as mangroves, coral reefs and seagrass beds, will face an uncertain future.

A less healthy future

Future changes to weather patterns may well prompt a number of adverse impacts on health. Direct influences will come from changes in the frequency of intense hot spells and cold snaps, regular extreme events such as floods and droughts, and fluctuations in air pollution. Indirect effects will come from changes to ecosystems altering the distribution of infectious diseases, levels of malnutrition and local food production.

If heat waves increase as forecast, the risk of death and serious illness will rise. Models of heat waves in urban areas predict that a number of US cities, for example, will experience several hundred extra deaths each summer. The death toll resulting from the extreme drought and heat wave that hit Europe in the summer of 2003 reached 35,000. Affecting mostly elderly people, it was one of the ten deadliest natural disasters in Europe in the last century and the worst in the last 50 years. The intense sunlight and stagnant air associated with heat waves helps to drive photochemical processes that create ozone at ground level, as well as causing fine particles such as soot and dust to hang in the air. The World Health Organization estimates that death rates go up by 0.3 per cent when ozone forms near the ground. British scientists concluded that between 21 and 38 per cent of UK deaths during the summer of 2003 could be classified as caused by ozone or particulate matter.

The increased intensity of extreme events such as storms, floods, droughts and cyclones may also result in loss of life and injury. Several major climate-related disasters have had adverse impacts on human health in recent years. These include floods in China, Bangladesh, Europe, Venezuela and Mozambique, as well as hurricanes such as Mitch and Katrina. Hurricane Katrina caused more than 1,800 deaths, not only because of the severity of the category 4 storm, with peak wind speeds of 249kmph, but because of the geography of New Orleans and the way in which local authorities responded to the disaster. Primarily below sea level, the city was protected by human-built levees, but these collapsed, allowing the sea to flood into intensely inhabited areas. The disaster response plan for New Orleans before Katrina struck was based on contending with a category 3 hurricane, with maximum wind speeds of 209kmph, so the authorities were not prepared for the intensity of the storm and the scale of the damage. The hurricane destroyed roads, bridges and causeways along with the communications infrastructure, making it difficult to access the dead and injured. People suffered from diarrhoeal illnesses, an outbreak of norovirus, respiratory infections and pneumonia.

Climate shifts such as higher temperatures and changes to rainfall will alter the distribution of infectious diseases spread by blood-feeding creatures such as ticks and mosquitoes. These include malaria, Lyme disease, dengue fever, leishmaniasis, various types of mosquito-borne encephalitis and tick-borne encephalitis. Presently, 40 per cent of the world lives in malarial locations; increased temperatures are liable to expand the range of the disease to higher latitudes and altitudes. Higher temperatures, in league with favourable patterns of rainfall and surface water, will also stretch the transmission season in some regions. However, increased temperatures and reduced rainfall may shrink the transmission range of those diseases that cannot tolerate warmer conditions. Overall, models suggest that the climate change scenarios over the coming century will increase the number of people living in transmission areas for malaria and dengue fever, and may boost the incidence of various water- and food-borne infectious diseases.

Changes in food supply resulting from climate change are likely to affect the nutrition and health of poorer sections of society. At present, some 790 million people are undernourished in the developing world, where the risk of reduced food yields is likely to be greatest. Other risks to health may result from changes to the marine environment increasing the occurrence of toxic algal blooms, such as one that washed up on Cornish shores in 2005, and "red tides" that have repeatedly affected tourist beaches in Florida. There are also some biotoxins, occurring in shellfish and fish eaten by humans, which thrive in warm tropical waters and which could increase their range to higher latitudes if the oceans were to warm up sufficiently.

Shifting plants and animals

At the simplest level, changing patterns of climate will alter the natural distribution limits of species and communities. Where there are no barriers to movement, some plants and animals will be able to migrate over time to new, favourable habitats. In the mid-latitude regions between 45° and 60°, present temperature zones could shift by between 150 and 550 kilometres. Several bird, butterfly and fish species are already on the move (see Chapter 2, page 84). Not all creatures will be able to up sticks and find a suitable new home, however. Many species are already threatened because their territories have become fragmented by human development. If climate change renders such isolated habitats intolerable, considerable numbers of plants and animals will be left with nowhere to go. Those that the World Conservation Union (IUCN) presently considers "critically endangered" will become extinct, while species that are "endangered" or "vulnerable" will become much rarer.

The 2004 Global Species Assessment was the most comprehensive evaluation of the world's biodiversity undertaken. It revealed that 15,589 plants, animals and lichens were at risk from extinction, including one in three amphibians, almost half of turtles and tortoises, one in eight birds and one in four mammals. This rate of extinction is one hundred to a thousand times higher than "natural" rates. Because only a fraction of known species have been assessed, the true number facing extinction is likely to be much higher. As well as the pressure from human development, overexploitation, competition from introduced species, pollution and disease have taken their toll on the world's biodiversity. Climate change will undoubtedly add to the pressures on the world's ecosystems, with unhappy outcomes for many of the species that inhabit them.

Some of the changes may mean that ecosystems that are today important habitats for wildlife will become much rarer or simply cease to exist. A temperature rise of 3°C or 4°C could eliminate 85 per cent of the world's wetlands. Sea-level rises will inundate coastal marshes, affecting both permanent inhabitants and migrating birds that are regular visitors. As these areas are often backed by urban developments or agricultural land, species squeezed out of coastal strips by the encroaching ocean will have no nearby suitable habitats to move to. Changes in rainfall and temperature will impact on forests, prompting more frequent fires and rapid changes to vegetation that outpace rates of migration or regrowth. Even relatively small temperature increases in the oceans have the potential to cause coral bleaching, and may kill off reefs altogether.

Biodiversity is linked to human well-being in many ways. Low-income families who rely on wildlife for subsistence will be directly affected as species dwindle. The reduced wealth of species will also affect the "eco-system services" that plants and animals provide, such as pollination and natural pest control. And loss of reefs and shifting ocean currents have implications for fisheries and aquaculture. Many species of plants and animals provide compounds that have medicinal uses for humans: compounds from foxgloves *Digitalis lanata* and *Digitalis purpurea* are used to treat heart failure; the Madagascar Periwinkle *Catharanthus roseus* provides two anti-tumour agents that are effective against childhood leukaemias; and the drug ASAQ, launched in March 2007, uses artemisin from *Artemisia annua* (Wormwood) to help prevent malaria. A quarter of the drugs used in the Western world are derived from rainforest species, yet scientists have scrutinized only a tiny percentage of plants living there. With climate change forecast to speed up already high rates of extinction, we stand to lose plants and animals that could be valuable to human welfare before we have even discovered them.

Changing holiday habits

Winter skiing trips and beach holidays in the Mediterranean could become pleasures of the past. The first International Conference on Climate Change and Tourism, held in Tunisia in 2003, sought to identify the major impacts of global warming. It concluded that seaside tourism would suffer from beach erosion, higher sea levels, damage from sea surges and storms and reduced water supply. Small island nations, which generate large percentages of their gross domestic product from tourism, are likely to be particularly badly affected. Several Caribbean islands were struck by Hurricane Ivan in 2004, with Grenada bearing the brunt of the 214kmph winds. During the peak tourism season five months after the event, hotels remained too badly damaged to be occupied. If small nations are battered more frequently by hurricanes and tropical storms, as some models predict, some may no longer be able to maintain their tourism industries.

In mountain regions, demand for winter sports is likely to drop as snow-covered areas shrink and snowfalls become less predictable. The skiing season will shorten, opportunities for young people to learn the sports will diminish, and pressures on fragile high-altitude environments will increase as resorts follow the receding snows up the mountains. Summer seasons could lengthen, increasing demand, but this may have further negative environmental consequences. The winter sports industry is already being affected by varying patterns of snowfall. Bolivia's only ski resort on the Chacaltaya glacier is likely to be snow-free in less than three years. Glaciers are increasingly melting across the Andes, but Chacaltaya's melting has been particularly speedy; the glacier has shrunk by 80 per cent in just 20 years. Bolivia once boasted an Olympic skiing team, but with artificial snow prohibitively expensive, the country's unique skiing industry may simply die out.

In Europe, the Alps are particularly sensitive to climate change, with recent warming roughly three times the global average. The years 1994, 2000, 2002 and 2003 were the warmest on record in the past 500 years. For the moment, 609 out of the 666 Alpine ski areas in Austria, France, Germany, Italy and Switzerland are considered as naturally "snow-reliable". The remaining nine per cent are already operating under marginal

Chacaltaya ski resort is being left high and dry by climate change.

conditions. The number of predictably snowy areas would drop to 500 under a 1°C temperature rise, to 404 under a 2°C warming, and to 202 under a 4°C increase. Farther north, the melting of snow in the Arctic could benefit tourism. A longer summer season might prove lucrative for cruise operators and ecotourism ventures offering activities such as wildlife-watching tours. However, Polar Bears and other creatures that now attract visitors may be much rarer or no longer exist.

Gloomy economic outlook

The varied effects of climate change could prompt a steep economic downturn if we do not move quickly to cap the world's spiralling temperature. Findings of the first major investigation into financial impacts, conducted by the former chief economist of the World Bank, Sir Nicholas Stern, were released at the end of October 2006 in *The Stern Review: The Economics of Climate Change*. Stern estimated that the costs of extreme weather alone could amount to between 0.5 and 1.0 per cent of the world's gross domestic product (GDP) per year by the middle of the century, and this figure would keep rising if the world continued to warm.

Stern's cost analyses of specific climatic events suggest that a five to 10 per cent increase in hurricane wind speed, linked to rising sea temperatures, would double annual damage costs in the USA. In the UK, annual flood losses could increase from today's level of 0.1 per cent of GDP to 0.2 per cent or 0.4 per cent if the rise in the average global temperature reaches 3°C or 4°C. Heat waves such as the one that struck Europe in 2003, when agricultural losses reached $15 billion, would be commonplace by the end of the century. Overall, Stern considered that a temperature rise of between 2°C and 3°C could reduce global economic outlook by three per cent. If average global temperatures rose by 5°C, up to 10 per cent of global output could be lost, but poorer countries would lose more than 10 per cent of their output. In the worst case scenario, he suggested, global consumption per head would plummet by 20 per cent.

The financial services sector, incorporating insurance and disaster relief, banking and asset management, is already feeling the heat from climate change. The costs of extreme weather events have risen in recent years; yearly global economic losses from large events increased from US$3.9 billion per year in the 1950s to US$40 billion per year in the 1990s. Around a quarter of these losses occurred in developing countries. The year 2005 was the most expensive ever for the insurance industry in relation to natural disasters, with overall losses amounting to US$210 billion and insured losses exceeding US$90 billion. A flurry of calamitous events killed more than 100,000 people. The hurricane season was particularly intense; for the first time since its introduction in 1953, the official list of 21 names chosen at the start of the season was not long enough to cover the 27 severe tropical cyclones that occurred in the Atlantic. Hurricane Katrina alone notched up US$125 billion in costs, making it the most expensive natural disaster in US history.

HOW THE WORLD WILL CHANGE IF IT CONTINUES TO WARM

+2°C impacts

HUMAN HEALTH Between 90 million and 200 million more people will be at risk of malaria and other vector- and water-borne diseases. More people in low-income countries will suffer from diarrhoeal disease and malnutrition.

AGRICULTURE A decline in agricultural production will lead to increased hunger in sub-Saharan Africa and south Asia. Canada, Russia and Scandinavia may benefit from a boost to crop yields but rapid rates of warming here could damage roads and buildings.

WATER Between 662 million and 3 billion more people will be at risk from water shortages. Dwindling global water supplies and parched soils will result in land being used more intensely, resulting in desertification.

ICE AND GLACIERS A 60 per cent loss of summer sea ice in the Arctic is likely. Antarctica's sea ice could decrease by 25 per cent. A 1.5°C increase in temperature could trigger melting of Greenland's ice sheet.

ECOSYSTEMS We stand to lose one-quarter of current species. Ninety-five per cent of most corals could perish by the middle of the century, with adverse impacts to subsistence and commercial fishing, tourism and the coastal protection provided by reefs. The associated economic cost could be A$4.3 billion (US$3.8 billion) per year for Australia's Great Barrier Reef alone. There will be a 43 per cent risk of change in global   forest to non-forest ecosystems, along with the expansion of forests into the Arctic and semi-arid savannas. Areas that have historically been stores of carbon, such as the Amazon rainforest and Arctic areas dominated by permafrost, could turn into sources instead. A major proportion of the tundra and about half of boreal forests may disappear.

SEA-LEVEL RISE Between 25 million and 50 million people will be at risk from sea-level rise and coastal flooding, costing nations hundreds of billions of dollars.

EXTREME WEATHER Increases in the frequency and intensity of floods, droughts, storms, heat waves, tropical cyclones, hurricanes and other extreme events will drive up economic costs and decrease opportunities for development.

MAIN SOURCE: WWF – CLIMATE CHANGE: WHY WE NEED TO TAKE ACTION NOW

+3°C impacts

HUMAN HEALTH More than 300 million more people will be at risk of malaria globally, and 5–6 billion more people could contract dengue fever. Human health will be threatened by water stress and flooding, especially in Africa and south Asia.

AGRICULTURE 50–120 million more people will face hunger. A decline in agriculture will cause food prices to increase globally.

WATER As many as 3.5 billion more people could be at risk of water shortages. Migrations caused by drought could lead to socio-economic and political instability. There will be a high risk of drought for southern Europe, West Africa, Central America, the Middle East, and parts of North America, Amazonia and China.

ICE AND GLACIERS Scientists expect the summer sea ice to almost completely disappear from the Arctic. A 3°C warming over several centuries will destroy the Greenland ice sheet and the Antarctic ice shelves.

ECOSYSTEMS We could lose around 33 per cent of current species. There will be little hope of recovery from annual bleaching of the remaining coral owing to the increased acidity of the oceans from abnormally high levels of absorbed CO_2. There is an 88 per cent risk that the global forests will change to non-forest systems, along with risks of forest losses in parts of Eurasia, Amazonia and Canada. Forests may also disappear from parts of the southern boreal zone, eastern China, Central America and the Gulf Coast of the USA. We face a much higher risk of terrestrial carbon sinks switching permanently to become carbon sources and irreversible damage to the Amazon forest leading to its collapse. Scientists anticipate the loss of half of wetlands in the Mediterranean and Baltic, along with several migratory bird habitats in the USA. Many ice-dependent species will likely become extinct, including Polar Bears.

SEA-LEVEL RISE 180 million people will be at risk of coastal flooding due to sea-level rise and water stress. Hundreds of thousands of people may have to migrate to other regions or countries.

EXTREME WEATHER Scientists expect massive increases in the frequency and intensity of fire, drought, storms and heat waves. Socio-economic losses from global damage could range from 3 to 5 per cent of GDP for developing countries, with a global average of 1–2 per cent for warming of between 2.5°C and 3°C.

+4°C impacts

HUMAN HEALTH   Mosquitoes will thrive, exposing 80 million more people to malaria in Africa; 2.5 billion more people are likely to become exposed to dengue fever.

AGRICULTURE   Droughts will cause African crop yields to slump by 15 to 35 per cent. Global food production could fall by 10 per cent.

WATER   The availability of freshwater will be halved in southern Africa and the Mediterranean.

ICE AND GLACIERS   Half the Arctic tundra will be at risk. Europe stands to lose 80 per cent of its Alpine glaciers. Melting of the West Antarctic and Greenland ice sheets will speed up.

ECOSYSTEMS   Half of land species may now be threatened with extinction.

SEA-LEVEL RISE   According to the IPCC's FAR, sea levels could rise by as much as 59cm. Bangladesh and Vietnam will be the worst hit, along with coastal cities such as London, New York, Tokyo, Hong Kong, Calcutta and Karachi. Small islands in the Caribbean and Pacific would become uninhabitable. There will be 1.8 million people at risk from coastal flooding in Britain alone.

EXTREME WEATHER: Hurricane wind strengths could increase by 15 to 25 per cent, causing great damage to buildings, roads and telecommunications infrastructure.

Mitigating the impacts of global warming

ACTION BEING TAKEN AT NATIONAL AND REGIONAL LEVELS

Switching to renewable energy sources such as wind power will help reduce our reliance on fossil fuels.

Levels of global CO_2 emissions are not only abnormally high, they are rising faster than ever, according to figures released by the Global Carbon Project, a consortium of international experts, in November 2006. Between 2000 and 2005, emissions grew four times faster than in the preceding decade. Growth rates were 0.8 per cent from 1990 to 1999 but rose to 3.2 per cent in the first five years of the new millennium.

This pattern of changing concentrations is beginning to follow the path of the IPCC's A1B scenario (as outlined on page 104 in chapter 3). This forecast assumes 50 per cent of the world's energy will still come from fossil fuels over the course of the next century. Current global concentrations of CO_2 stand at 379 parts per million (ppm) (against 280ppm around 1900). If we continue along this carbon emissions path, we are likely to see CO_2 levels rise to at least 550ppm.

Similarly, the planet's temperature has been climbing more rapidly in recent years. The rise has been most pronounced since 1976; in the past 30 years the temperature has risen by 0.18°C per decade. The year 2006 was the fifth warmest year on record globally (NASA) and the UK's warmest year since its records began in 1659. Almost all parts of the world experienced some extreme weather events, suggesting that the temperature rise is beginning to have a serious impact on the global climate. Brazil and Australia experienced heat waves between January and March, with one Brazilian town, Bom Jesus, notching up 44.6°C on 31 January, one of the highest temperatures recorded in the country. Flooding affected the entire Great Horn of Africa, with some regions receiving more than six times their average monthly rainfall. Even the Sahara was not spared; floods in February damaged 70 per cent of food production and displaced 600,000 people. And in China, millions of hectares of crops shrivelled as droughts dried up Sichuan province, and the country suffered its worst typhoon season in a decade. Meanwhile, the Arctic sea ice melted away during the summer at the unprecedented rate of 60,421 km^2 of sea ice per year.

So what, if anything, is being done to halt the change? The subject of climate change entered the political arena in 1994, when the UN Framework Convention on Climate Change (UNFCCC) came into force, but progress towards making any substantial cuts in greenhouse gas emissions has been slow. Though its aim at the outset was "to achieve stabilization of greenhouse gas concentrations in the atmosphere at a low enough level to prevent dangerous anthropogenic interference with the climate system", the framework set no mandatory limits on greenhouse gas emissions for individual nations and was legally non-binding. Instead, it included provisions for updates, called "protocols", which would set mandatory emission limits. The main update is the Kyoto Protocol, which came into effect in February 2005 and has set targets for emissions of six greenhouse gases: carbon dioxide (CO_2), methane (CH_4), nitrous oxide (N_2O), sulphur hexafluoride (SF_6), hydrofluorocarbons (HFCs) and perfluorcarbons (PFCs). The targets range from –8 per cent to +10 per cent of the country's individual 1990 emissions levels "with a view to reducing their overall emissions of such gases by at least five per cent below existing 1990 levels in the commitment period 2008–2012".

In all, 189 countries have ratified the UNFCCC, nearly all the countries in the world. This is important, as only parties to the UNFCCC can sign or ratify the Kyoto Protocol. As of December 2006, 169 countries and other governmental entities had ratified the Kyoto Protocol, but notable in their

Former Canadian
Prime Minister, Jean
Chrétien, signing the
Kyoto Protocol
in 2002.

absence were the USA (the world's biggest emitter of greenhouse gases) and Australia. Parties are classified as Annex I or Non-Annex I nations. The former comprise either Annex II countries, representing the world's most developed countries, or those with Economies in Transition (EIT), which are mostly from Eastern Europe and Russia. Non-Annex I (NAI) countries, primarily developing nations, make up the remainder. Annex II Parties are expected to support developing countries financially, so they can undertake schemes to reduce emissions and adapt to adverse effects of climate change. Developing economies are currently exempt from having to cut emissions. This is only fair, as their per capita emissions are still relatively low and developed countries contributed most of the current greenhouse overload. However, because India and China are developing rapidly, they are likely to become the top contributors of greenhouse gas emissions in the near future.

The overall global cut of five per cent is designed to be achieved through cuts of eight per cent in the European Union (EU-15), Switzerland and most central and East European states; seven per cent in the USA (which withdrew its support for the Protocol); and six per cent in Canada, Hungary, Japan and Poland. New Zealand, Russia and Ukraine are supposed to work towards stabilizing their emissions, while Norway is permitted a one per cent rise. Australia (which also withdrew its support) is permitted an eight per cent increase. The agreement offers some flexibility in terms of how countries meet their targets. For example, they can partly offset emissions by creating "carbon sinks". The UNFCCC defines a "sink" as "any process, activity or mechanism which removes a greenhouse gas, an aerosol or a precursor of a greenhouse gas from the atmosphere". Called "flexibility mechanisms", these include schemes to plant carbon-absorbing forests and other activities that result in greenhouse gas cuts. (See page 134 for more information on Kyoto's flexibility mechanisms.)

Each year, the Annex I countries submit their greenhouse gas emissions data to the UNFCCC secretariat, which publishes an annual report. This includes greenhouse gas data for Annex I countries that have ratified the Kyoto Protocol as well as those that have not. Key findings of the 2006 report are that greenhouse gas emissions fell by 3.3 per cent between 1990 and 2004. While this sounds relatively positive, given the five per cent target for 2008–2012, it breaks down to a 36.8 decrease for EIT and an

11 per cent increase for non-EIT Parties. In other words, the industrialized world's increases in greenhouse gas use have been offset by the massive decline in industry in Eastern European and former Soviet countries following the collapse of communism. In all, emissions from 22 Annex I Parties decreased over this period, while the output of 19 countries increased. Since 2000, emissions from both EIT and non-EIT Parties have increased slightly and the number of parties with emission decreases has dropped. Between 1990 and 2000, some 23 of the 41 Annex I countries reported lower emissions, but between 2000 and 2004 only seven countries registered decreases. Changes in greenhouse gas emissions between 2000 and 2004 vary wildly, from Lithuania's 60.4 per cent decrease to Turkey's 72.6 per cent increase. Some scientists believe certain governments are reporting incorrect emissions figures and have called for independent audits of measurements (see box Greenhouse gas omissions on page 133).

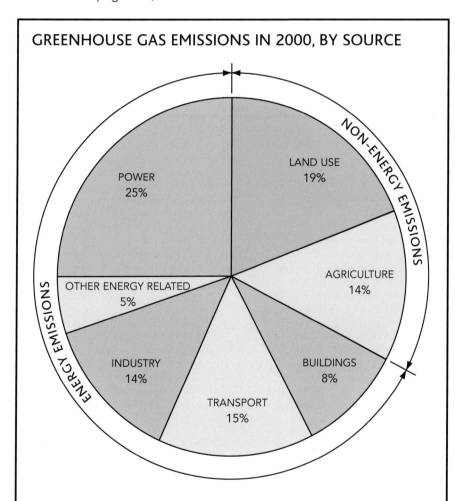

GREENHOUSE GAS EMISSIONS IN 2000, BY SOURCE

NON-ENERGY EMISSIONS

LAND USE
19%

POWER
25%

AGRICULTURE
14%

OTHER ENERGY RELATED
5%

INDUSTRY
14%

BUILDINGS
8%

TRANSPORT
15%

ENERGY EMISSIONS

Energy emissions are mostly CO_2 (some non-CO_2 in industry and other energy-related).
Non-energy emissions are CO_2 (land-use) and non-CO_2 (agriculture).

Diagram showing the breakdown of carbon emissions from different industry sectors for the year 2000.

The aviation industry
is the fastest
growing contributor
to greenhouse
gas emissions.

Some sectors have managed to reduce emissions in recent years. The greatest decrease between 1990 and 2004 was in agriculture (–20 per cent), but energy industries and transport posted large rises of +8.6 per cent and +23.9 per cent. The report stated that: "In all EIT Parties except Slovenia, the emissions are now much below the targets under the Kyoto Protocol. For the non-EIT Parties, some Parties (such as France, Germany, Greece, Iceland, Monaco, Sweden and the United Kingdom of Great Britain and Northern Ireland) are currently relatively close to their targets; other Parties require substantial additional reductions or need to use the international flexibility mechanisms under the Kyoto Protocol". It concluded: "Industrialized countries will need to intensify their efforts to reduce greenhouse gas emissions. Transport remains a sector where emissions reductions are needed but seem to be especially difficult to achieve".

The biggest worry as far as transport emissions are concerned comes from aeroplanes. The aviation industry is the fastest growing contributor to greenhouse gas emissions. It is currently responsible for around 3.5 per cent of all human-cause climate change emissions, but this figure is forecast to rise to 15 per cent of the worldwide total by 2050. Greenhouse gas emissions from fuels sold for international aviation rocketed by 52 per cent between 1990 and 2004. According to Friends of the Earth, the world's 16,000 commercial jet aircraft generate more than 600 million tonnes of CO_2 per year. Rapid development in India and China is contributing greatly to the industry's expansion. It is likely that there will be 70 million air passengers in India alone by 2010. Aeroplane manufacturer Boeing has said it expects India will need 856 new jet aircraft over the next 20 years; the domestic market alone is likely to expand by 20 per cent over the next five years. Competition among new airlines has resulted in airfares challenging rail fares, and has opened up flying to a much wider range of people. And, for the time being, India has no obligation to reduce its carbon emissions under Kyoto.

GREENHOUSE GAS OMISSIONS

A report in the June 2006 edition of the weekly magazine *New Scientist* suggests that governments are not being entirely truthful about their greenhouse gas emissions data. Under Kyoto, each government works out how much carbon dioxide, methane and nitrous oxide its country emits by adding together estimated emissions from individual sources. However, the measurements have never been independently audited. The article reported that two independent studies both found that certain countries were under-reporting emissions of methane.

The first, by Peter Bergamaschi of the European Commission Joint Research Centre at Ispra, Italy, used a "top-down" technique which involved studying variations in levels of greenhouse gases around the world. By tracking air movements and measuring the differences in gases present over built-up areas and oceans, scientists say they can deduce a country's emissions independently of government estimates. According to Bergamaschi's calculations, the UK emitted 4.21 million tonnes of methane in 2004, almost double the 2.19 million tonnes it declared, and France emitted 4.43 million tonnes, against the 3.10 million tonnes it reported. Both countries stick by their original estimates, but during Bergamaschi's study, the German Government revised its estimate of national methane emissions upwards by some 70 per cent, placing it close to his estimate.

The second study was led by Euan Nisbet of Royal Holloway, University of London. A member of the Global Atmosphere Watch (GAW), a network of atmospheric scientists organized by the UN's World Meteorological Organization, Nisbet estimates that methane emissions in the London area in the late 1990s were 40 to 80 per cent higher than declared by the Government at the time.

Nisbet, Bergamaschi and other scientists are calling for a global monitoring system that will allow external researchers to audit emissions claims by directly measuring concentrations of greenhouse gases in the air. The present monitoring network is far from comprehensive; China and India have only one monitoring site each, and the European Union recently shut down its continent-wide methane-monitoring programme. Only when a new network is set up will scientists be able to independently calculate who is emitting what and expose countries that are not complying with their obligations under the Kyoto Protocol.

Flexibility mechanisms

Under the Kyoto Protocol, countries can use one of three flexibility mechanisms to help them reach their emissions targets. They are Joint Implementation (JI), the Clean Development Mechanism (CDM), and Emissions Trading. However, the countries that do so have to prove that their use of the mechanisms supplements significant efforts being made at home to meet their commitments. Ironically, the mechanisms were included largely on the USA's insistence, before the world's biggest polluter withdrew its support for the Protocol. The carbon credits allocated to governments with Kyoto targets are called Assigned Amount Units (AAUs). In addition, there are three other types of credit, called Certified Emission Reductions (CER), Emission Reduction Units (ERU), and Removal Units (RU), which are generated through the different types of project. Point Carbon, the Norway-based carbon advisory house, predicts the global trade in carbon will reach €34 billion (US$45 billion) by 2010, with some 4.5 billion tonnes of CO_2 or equivalent being traded in allowances or through JI and CDM schemes.

Joint Implementation enables developed nations to invest in emissions-reducing activities in other industrialized countries. This is beneficial if action taken on their home turf would be more costly. Projects gain ERUs, which they can put towards their national emissions targets or to sell on the international emissions trading market. An example is the Sawdust 2000 Project, launched by the Danish Environmental Protection Agency and the Romanian Ministry of Environment and Water Management in 2001. The aim of the €14 million (US$20 million) project was to use waste sawdust from sawmills in place of fossil fuels to power district heating systems. After a pilot project in the small Romanian town of Tasca in 1999–2001, five plants began operating across the country in 2004. The intention is generate 700,000 ERUs and AAUs during the period 2004–2012.

The Clean Development Mechanism allows industrialized nations to invest in emissions-slashing schemes in developing countries. In doing so they gain CERs, which count towards Kyoto targets. The aim is to help boost developing countries' sustainable development. As of September 2006, the UK was participating in 124 CDM schemes. Most recently approved were 31 small-scale projects aimed at generating electricity from pig waste in Mexico. The plan is to use methane gas recovered from waste at piggeries to mitigate 310,000 tonnes of CO_2 equivalent (the "carbon dioxide equivalent" measure is a way of considering the potency and lifetime of all greenhouse gas emissions in relation to CO_2) every year by reducing reliance on the Mexican electricity grid. Other projects with UK involvement include a 30-megawatt wind farm project in China, a hydroelectric project and a scheme to boost energy efficiency at a hotel in India, and a geothermal project in Indonesia.

In theory, the CDM has the potential to benefit developing nations by bringing in much-needed foreign investment. However, because companies are allowed to work towards reducing any of the six greenhouse gases defined under Kyoto, and the cheapest reductions are to be had in cutting HFCs, almost a third of projects are aimed at reducing the refrigerator gas HFC-23. Only three per cent are CO_2-lowering projects. In the industrialized world, cutting HFCs is so cheap that manufacturers have opted to cut their emissions voluntarily. Yet abroad, HFC-23 emitters can earn almost twice as much from CDM credits as they can from selling refrigerant gases. In terms of the amount of emissions reduced per pound sent, this makes the CDM a very inefficient subsidy.

Other criticisms are that enabling industrialized countries to use land in developing nations may lead to a form of "carbon colonialism", in which projects exploit local people and degrade their natural environment. One such case has been uncovered in eastern Africa. The ethical news agency, Panos, reported that the Uganda Wildlife Authority drove indigenous Ndorobo families from their ancestral lands in Mount Elgon National Park before signing a 25,000-hectare tree plantation agreement with a Netherlands-based carbon trading company. In retaliation, Ndorobo men hacked down thousands of trees planted by the company. Similar stories of violence and displacement have emerged from Kibale National Park in western Uganda. As well as causing social unrest, tree plantations can degrade the soil, making it unsuitable for food crops in the future.

The third flexibility mechanism is Emissions Trading. This allows nations whose carbon emissions are below their Kyoto target to sell carbon credits to countries whose emissions exceed their targets. The largest carbon trading scheme is that set up by the European Union in an effort to meet its eight per cent Kyoto target. The EU Emissions Trading Scheme (ETS) is a cap-and-trade system which aims to limit carbon emissions from large industrial sources. The scheme works by allocating greenhouse gas emission allowances, where each allowance represents one tonne of CO_2 equivalent. However, for the first phase of the scheme, which runs from 1 January 2005 to 31 December 2007, CO_2 is the only gas covered by the EU ETS. Each member state sets an overall limit, or cap, on the total number of allowances to issue to installations, based on its Kyoto target. Industries including power generation, iron and steel, glass and cement and paper receive an allocation of allowances. Each year, they have to provide emissions data for the preceding year and surrender the required number of allowances. The idea is that those operations that are able to reduce emissions at a low cost can do so and then sell their surplus allowances to installations for which reducing emissions would be more expensive. The scheme covers some 12,000 energy-intensive plants, which spew out around 40 per cent of the EU's total CO_2 emissions.

Carbon analyst Point Carbon estimates that in 2005 the EU ETS traded some 362 million tonnes of CO_2, with an estimated financial value of 7.2 billion Euros. In 2006, the European Commission proposed that aviation should also be included within the trading scheme from 2011. Aircraft currently emit more than refineries or steel plants and CO_2 outputs from the sector have risen by 87 per cent since 1990 with the emergence of cheap air travel. The Commission said that 46 per cent of the expected growth in aviation emissions – 183 million tonnes of CO_2 per year – would be saved if its plan were implemented in full. As with other industries, the plan would work by issuing airlines with emissions allowances, mostly free of charge, based on their average carbon use between 2004 and 2006. However, much of the anticipated "saving" would come through other participants of the scheme selling unused allowances to the airlines, rather than cuts directly from aviation.

The EU's trading scheme is widely accepted as a positive step towards cutting down emissions and reducing climate change. However, nations issued more permits than were needed in the first phase. The result has been that carbon prices have dropped as low as €8 (£5.40) per tonne, so that it has been much cheaper for firms to buy surplus credits than to make the required cuts or pay the €40 (£27) fine that is levied for every tonne of CO_2 they produce in excess of their allocation. As this book was being written, talks were under way to agree national allocations for the second phase of the scheme, between 2008 and 2012. In January 2007, the European Commission set new targets for EU greenhouse gas emissions at 20 per cent below 1990 levels by 2020. The ETS will play a key role in helping to deliver this target.

Europe's emissions trading scheme is not the only one in existence. The Chicago Climate Exchange (CCX) was developed in the early 2000s. Its members, which include the city of Chicago and Mexico City, make voluntary but legally binding commitments to reduce their emissions. In the eight months ended 31 August 2006, CCX traded approximately eight million tonnes of carbon, a considerable increase on the 0.53 million tonnes it traded during the same period in 2005. When a similar cap-and-trade emissions trading scheme became law in California in January 2007, CIBC World Markets predicted that, by the end of the decade, every North American state and province would follow its lead on carbon emission laws. Ultimately, there is likely to be a global emissions trading scheme put in place.

Emissions trading has its critics, however. Many environmentalists believe that applying free-market principles to the natural environment is a false solution that simply enables countries and companies to trade the "rights to pollute". They believe the only acceptable option is to leave fossil fuels in the ground and live more sustainable lives. The Durban Group for Climate Justice is an international network of independent organizations, individuals and people's movements who believe that excessive burning of fossil fuels is jeopardizing Earth's ability to maintain a liveable climate, and reject the free market approach to climate change. In 2004 it published the Durban Declaration on Carbon Trading, denouncing "the further delays in ending fossil-fuel extraction that are being caused by corporate, government and United Nations' attempts to construct a 'carbon market', including a market trading in 'carbon sinks'".

POLITICS OF CLIMATE CHANGE TIMELINE

1979: First World Climate Conference explores how climate change might affect human activities and issues a declaration calling on the world's governments "to foresee and prevent potential man-made changes in climate that might be adverse to the well-being of humanity". It also endorses plans for a World Climate Programme (WCP) under the joint responsibility of the World Meteorological Organization (WMO), the United Nations Environment Programme (UNEP), and the International Council of Scientific Unions (ICSU).

Late 1980s and early 1990s: Several intergovernmental conferences take place, focusing on climate change, and calling for global action.

1988: The Intergovernmental Panel on Climate Change (IPCC) is set up by UNEP and WMO to assess: the state of existing knowledge about the climate system and climate change; the environmental, economic, and social impacts of climate change; and possible response strategies.

1990: The IPCC releases its first assessment report in which it states, "We are certain of the following: ... emissions resulting from human activities are substantially increasing the atmospheric concentrations of the greenhouse gases: CO_2, methane, CFCs and nitrous oxide. These increases will enhance the greenhouse effect, resulting on average in an additional warming of the Earth's surface." However, it concludes that "the unequivocal detection of the enhanced greenhouse effect is not likely for a decade or more".

The Second World Climate Conference calls for the development of a framework treaty on climate change. In December, the UN General Assembly approves the start of negotiations for a framework convention on climate change.

1991–1992: The newly formed Intergovernmental Negotiating Committee (INC) meets for five sessions. Within 15 months, negotiators from 150 countries finalize the convention. It is approved in New York on 9 May 1992 in time for the Rio de Janeiro Earth Summit in June, at which 154 nations sign up to it. Its stated objective is "to achieve stabilization of greenhouse gas concentrations in the atmosphere at a low enough level to prevent dangerous anthropogenic interference with the climate system".

1994: Having been ratified by more than 50 countries, the United Nations Framework Convention on Climate Change (UNFCCC) comes into force. However the member countries soon decide it needs to be augmented by an agreement with stricter demands for reducing emissions.

1995: The INC is dissolved after its eleventh and final session in February and the Conference of the Parties (COP) becomes the Convention's new authority. Negotiations for a protocol to support the UNFCCC get under way. The COP's first session launches the Berlin Mandate talks on new commitments. The IPCC adopts its Second Assessment Report, written and reviewed by 2,000 scientists worldwide. It concludes that "the balance of evidence suggests that there is a discernible human influence on global climate".

1996: The Second Conference of Parties to the UNFCCC (COP-2) formally accepts the findings of the IPCC's Second Assessment Report.

1997: COP-3 is held in Japan, and adopts the Kyoto Protocol to the UNFCCC. Most industrialized nations and some Central European countries with "economies in transition" agree to legally-binding reductions in greenhouse gas emissions of an average of six to eight per cent below 1990 levels between the years 2008 and 2012.

1998: COP-4 takes place in Buenos Aires, Argentina. The participants adopt a two-year plan of action to thrash out the finer details of how the Kyoto Protocol will operate.

1999: COP-5 in Bonn, Germany, does not draw any major conclusions.

2000: COP-6 is held in The Hague, but the meeting is suspended when negotiations over the major political issues fail to reach agreement.

2001: COP-6 negotiations resume in Bonn in July. Since the previous meeting, George Bush has taken over as President from Bill Clinton and officially rejected the USA's ratification of the Kyoto Protocol. As a result, the USA delegates to the meeting act as observers only. The remaining Parties reach agreement on most of the outstanding major issues. These include allowing flexible mechanisms, such as Emissions Trading, Joint Implementation and the Clean Development Mechanism.

At the COP-7 meeting that takes place in Marrakesh, Morocco, the same year, negotiators complete the Plan of Action outlined at COP-4. The Protocol is finally ready to be implemented.

The IPCC Third Assessment Report states that global warming, unprecedented since the end of the last ice age, is "very likely", with possible severe surprises.

2002: COP-8 takes place in New Delhi, India.

2003: COP-9 takes place in Milan, Italy.

2004: To put the Kyoto Protocol into effect requires ratification by nations with more than 55 per cent of the world's CO_2 emissions. With the USA refusing to join, only Russia can put the treaty into effect. The Russian Government finally ratifies the treaty under pressure from Europe. COP-10 later takes place in Buenos Aires, Argentina.

2005: In February 2005 the Kyoto Protocol comes into effect with 141 signatory nations. It is generally acknowledged as a first step towards putting in place systems to monitor and control emissions, setting up carbon trading schemes and stimulating the invention and development of energy-saving technologies and practices. A new treaty is anticipated for when the Kyoto Protocol reaches its end in 2012.

In June, the science academies of the world's leading industrial and developing countries sign an unprecedented joint statement, declaring that "the threat of climate change is clear and increasing", and calling on all nations to take "prompt action to reduce the causes of climate change, adapt to its impacts and ensure that the issue is included in all relevant national and international strategies".

In November, COP-11 takes place in conjunction with the first Meeting of the Parties to the Kyoto Protocol (MOP-1) in Montreal, Canada. Delegates adopt the so-called Marrakesh Accords, the rule-book for the Kyoto Protocol, formally making it operational. They also discuss the details for emissions trading mechanisms. On the last day of the conference, signatories agree to extend the treaty on emissions reductions beyond its 2012 deadline, while a broader group of countries, including the USA, agree to non-binding talks on long-term measures.

2006: In November, Nairobi hosts MOP-2 in conjunction with the COP-12. It is the first conference of its kind in sub-Saharan Africa and the largest environmental conference held in Kenya, drawing up to 7,000 participants. The talks end with agreement reached on all outstanding matters. However, there is no deal on another round of mandatory cuts in greenhouse emissions when the current Kyoto quotas expire in 2012, and no firm time-frame for agreeing new targets. David Miliband, the then UK Secretary of State for the Environment, Food and Rural Affairs, later admits: "The gap between what the science tells us is necessary and what the politics is delivering is still significant".

2007: The IPCC releases the first part of its Fourth Assessment Report. The panel confirms the link between climate change and human activity, saying that it is at least 90 per cent certain that human emissions of greenhouse gases are warming the planet's surface. Talks are planned for December in Bali, Indonesia.

Avoiding dangerous climate change

As well as aiming "to achieve stabilization of greenhouse gas concentrations in the atmosphere at a low enough level to prevent dangerous anthropogenic interference with the climate system", the UNFCCC intended that such a level should be achieved with time to allow ecosystems to adapt naturally to climate change, ensure food production is not threatened, and enable economic development to advance in a sustainable way. But with Kyoto failing to deliver deep cuts, and developing world countries increasing their emissions, the question is, how high can greenhouse gas concentrations rise before the threshold defining dangerous climate change is crossed? In 2005 the UK Government sought to answer this question by convening a conference on Avoiding Dangerous Climate Change at the Met Office in Exeter, UK. Specifically, it wanted the gathering of esteemed scientists from around the world to determine "What level of greenhouse gases in the atmosphere is self-evidently too much?" and "What options do we have to avoid such levels?".

To answer these questions, you need to know what constitutes "dangerous climate change". One definition emerged at the European Climate Forum's Symposium on Key Vulnerable Regions and Climate Change in Beijing in 2004. It outlined "determinative dangers" as including "circumstances that could lead to global or unprecedented consequences, extinction of iconic species or loss of entire ecosystems, loss of human cultures, water resource threats and substantial increases in mortality levels, among others". The levels of danger associated with particular climatic circumstances can be quantified through "key vulnerabilities". For humans these include monetary loss, death, hunger, risk of water shortages, and the number of people threatened by coastal flooding or malaria infection, or forced to migrate because of the impacts of climate change. For Earth's natural systems, they include the number of species lost through extinction, the extent of changes to habitats, and the thresholds of the world's "tipping points", such as the shutdown of the Meridional Overturning Circulation, disintegration of the West Antarctic ice sheet, and widespread bleaching of coral reefs.

The heart of the debate at the Met Office conference centred on the temperatures above which dangerous climatic events might take place. The IPCC's 2001 Third Assessment Report (TAR), the most up-to-date report at the time, estimated that by the year 2100 global average surface temperatures would rise by between 1.4°C and 5.8°C. Scientists considered that warming at the low end of this scale would be relatively less stressful, but could still have a significant effect on some "unique and valuable systems". But warming at the high end of this range would have widespread catastrophic consequences, as a temperature change of 5°C to 7°C on a globally averaged basis is roughly the difference you would get between an ice age and an interglacial period. The European Union's greenhouse gas reduction target, set in 1996, aimed to prevent a rise in global average temperature of more than 2°C. However, the conference concluded that even that increase could be too high, as a global rise of just 1.5°C may be enough to trigger melting of the Greenland ice sheet. This would have a major impact on sea levels globally, although it would take up to 1,000 years to see the full predicted rise of seven metres. Above two degrees, the participants concluded, the risks increase "very substantially", with "potentially large numbers of extinctions" and "major increases in hunger and water shortage risks ... particularly in developing countries".

The scientists also calculated what concentrations of CO_2 in the atmosphere would be enough to cause such "dangerous" temperature increases. They suggested that levels would need to be stabilized at 400ppm to be in with a good chance of keeping the temperature rise below 2°C. However, at the current rate of greenhouse gas emissions, we are likely to reach 400ppm in just a decade. And stabilizing at 450ppm is looking increasingly unrealistic. In October 2006, environmental campaigner WWF released the report *Climate change: Why we need to act now*, which looked at the potential difference in impacts of 2°C and 3°C of warming. It concluded that an "average global warming of 2°C will result in dangerous and irreversible effects, which rapidly worsen above 2°C warming". For example, with a 2°C warming, 90–200 million more people will be at risk from malaria, but at 3°C this rises to more than 300 million. At 2°C, 25–50 million people will be at risk from coastal flooding because of sea-level rise, but at 3°C this jumps to 180 million. And at 2°C, we will lose 60 per cent of the summer sea ice in the Arctic, but at 3°C nearly all the summer ice will disappear.

In late 2005, Jim Hansen, head of NASA's Goddard Institute for Space Studies, gave a presentation at the American Geophysical Union, entitled "Is There Still Time to Avoid 'Dangerous Anthropogenic Interference' with Global Climate?". In it he suggested that a more appropriate level for what constitutes dangerous climate change is even lower, at 1°C. He pointed out that, during the past several hundred thousand years, the global temperature reached a temperature only 1°C warmer than today, but at the peak of the last interglacial, sea level was as much as six metres higher. To find a time when Earth's average temperature was 2°C or 3°C higher, we have to go back three million years, to within the Neogene (see Earth history timeline, page 184). And at that time, sea level was 25 metres (plus or minus ten metres) higher than today. If ice sheets respond to higher temperatures on timescales of centuries, and we carry on pumping out greenhouse gases along "business as usual" lines, he said, total sea-level rise might be a couple of metres this century and several more next century. And the changes could be unpredictable; while snow builds up in a dry, linear way, ice sheets can disintegrate in rapid pulses, with associated leaps in sea level. According to Hansen, if we do not act to change technologies and attitudes within a decade, China, India and the USA will bring planned coal-fired power stations into operation with no integral carbon sequestration technology. This alone may prevent us keeping global warming under 1°C.

Greenhouse gas emissions from transport in Annex I countries increased by 23.9 per cent between 1990 and 2004.

World leaders who attended the last two G8 summits, at Gleneagles in Scotland and St Petersburg in Russia, called on the International Energy Agency (IEA) to advise on how the world might reduce its reliance on fossil fuels. In response, it published *World Energy Outlook 2006* in December 2006. The report put forward two scenarios for how the world's energy markets might look by 2030. The first showed how markets were likely to evolve if no new government measures were put in place to alter current energy trends. This anticipated a doubling in the demand for energy around the world, with 70 per cent of the increase coming from developing countries, led by China and India. In this world, the demand for oil would rise, from 84 million barrels a day (mb/d) to 116 mb/d in 2030. As a result of this continued heavy use of fossil fuels, the report predicted, CO_2 would increase by 55 per cent. China would overtake the USA as the world's biggest emitter as early as 2010. And the magnitude of global climate change would be greatly increased as a result of this "business as usual" path.

The IEA's alternative scenario simulated the effect of 1,400 policies to reduce the use of fossil fuels. It demonstrated that the world's energy future could be improved if governments implemented measures they are currently considering. But it also underlined the sheer size of the task ahead. Taking into consideration large improvements in efficiency, an increased use of nuclear power and a quadrupling of the global share of energy from renewables to eight per cent, the anticipated outcome was a rise in CO_2 emissions by 2030 of 31 per cent, rather than 55 per cent. And even attaining this level of independence from fossil fuels does not come cheap. The report predicted that quenching the world's current thirst for energy would

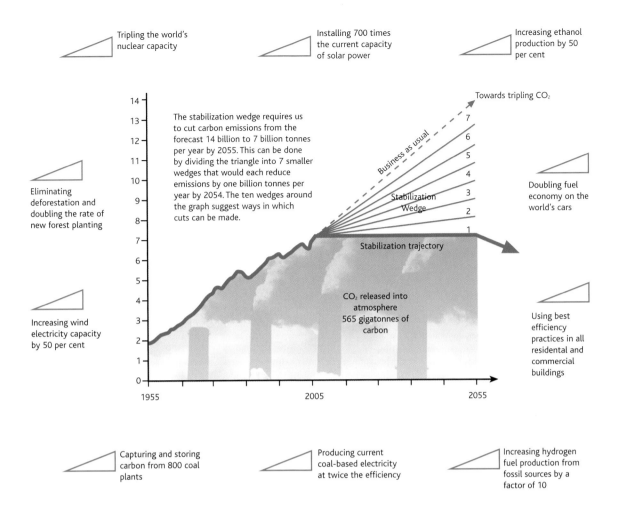

Tripling the world's nuclear capacity

Installing 700 times the current capacity of solar power

Increasing ethanol production by 50 per cent

Towards tripling CO₂

The stabilization wedge requires us to cut carbon emissions from the forecast 14 billion to 7 billion tonnes per year by 2055. This can be done by dividing the triangle into 7 smaller wedges that would each reduce emissions by one billion tonnes per year by 2054. The ten wedges around the graph suggest ways in which cuts can be made.

Business as usual

Stabilization Wedge

Eliminating deforestation and doubling the rate of new forest planting

Doubling fuel economy on the world's cars

Stabilization trajectory

CO₂ released into atmosphere 565 gigatonnes of carbon

Increasing wind electricity capacity by 50 per cent

Using best efficiency practices in all residental and commercial buildings

1955   2005   2055

Capturing and storing carbon from 800 coal plants

Producing current coal-based electricity at twice the efficiency

Increasing hydrogen fuel production from fossil sources by a factor of 10

require investment between 2005 and 2030 of some US$20 trillion, around half of which would be needed in developing countries. On a more positive note, a report published by the American Solar Society in February 2007 claims to have found a way to cut the USA's CO₂ emissions by 1.2 billion tonnes annually by 2030, a figure cited as necessary to prevent the most dangerous consequences of climate change. The cuts would be achieved by boosting energy efficiency in buildings, transportation and industry, and using renewable technologies including concentrating solar power, photo-voltaics, wind power, biomass, biofuels and geothermal power.

The question is whether our current economic set-up, based on consumerism, can deliver us safely to a low-carbon future. With the Kyoto Protocol having failed so far to deliver substantial cuts in greenhouse gases, many people believe the global system for switching to a low-carbon economy needs to be rethought. A better, fairer method, which has gained wide support among scientists and policymakers, is one of contraction and convergence, developed by the Global Commons Institute. "Contraction" involves cutting the world's output of greenhouse gases, and "convergence" entails sharing out among all countries the amount of climate pollution which scientists say the Earth can tolerate. An emissions "ceiling" is set, fixing the future global emissions budget. Ideally

We may be able to cut emissions of CO₂ by spreading the load across a portfolio of technologies.

this would be no greater than 450ppm, to keep us from crossing the 2°C threshold, but it may already be too late. The budget would shrink year on year until reaching zero by around 2080. The tradeable shares in this future budget would be allocated as "one person, one share" globally. However, this would be moderated by a convergence to the global average of equal per capita shares over 20 or 30 years. If the world took this route, by around 2050 everyone in the world would be entitled to emit the same quantity of greenhouse gases, be they wealthy Westerner or African villager.

Annual global emissions of carbon from CO_2 are presently 7.2 gigatonnes and are rising by two per cent a year. To stabilize concentrations of the gas over the next few years will require cuts of at least 60 per cent. But because of CO_2 stored in the oceans, long-term stabilization will actually demand cuts of between 60 and 80 per cent. One practical method that has been put forward for tackling this monumental task is to divide the problem up into smaller goals. The idea was developed by Stephen Pacala and Robert Socolow of Princeton University, USA, and published in *Science* in 2004. The authors suggest that, although no single technology exists that can single-handedly deliver the required deep cuts in CO_2, spreading the load across a portfolio of proven technologies could achieve stabilization. They advocate dividing the segment separating the upward "business as usual" path from the horizontal "stabilization path" into a series of smaller wedges (see diagram on page 143). Seven wedges, each representing a billion tonnes (a gigatonne) of carbon from CO_2, would each be assigned a different technology or strategy. Pacala and Socolow have identified 15 possible wedges, representing technologies or approaches that they say are already being implemented somewhere in the world, although scaling up from the current capacity to the size required to eliminate a wedge's worth of CO_2 could present a significant challenge.

Most scientists are under no illusions that meeting the challenge posed by climate change will be easy. Keeping further global warming below the danger threshold will require an initial flattening out and then decreasing of CO_2 emissions, and an absolute decrease in non-CO_2 greenhouse emissions, particularly methane and carbon monoxide (cutting tropospheric ozone) and black carbon (soot) aerosols. Massive injections of funding will be needed to support research into non-fossil and carbon sequestration technologies, along with political support and strong leadership from the developed and developing world alike. As Marty Hoffert, Professor Emeritus of Physics at New York University has observed, "Technologies to fix global warming exist in the sense that the ability to make a nuclear weapon existed in the late 1930s, or the ability to send a crewed exploration craft to the Moon and return existed in the late 1950s. But it took the Manhattan and Apollo programs to make them so". The alternative – adaptation – is no alternative, since action such as building sea walls, installing air-conditioning and irrigating previously rain-fed fields will also demand huge amounts of energy. The belief still exists that the human race is up to the greatest challenge it has ever faced, but time is fast slipping away.

The world's biggest emitters and their progress towards reducing greenhouse gas emissions

USA

Population: 293.95 million

2004 CO_2 emissions: 5,799.97 million tonnes of CO_2

Emission reduction target under Kyoto: 7 per cent but USA has rejected Protocol

Changes in emissions 1990–2004: +15.8

The USA emits more, absolutely and per head, than any other country; the average US citizen emits as much CO_2 in one day as someone in China does in more than a week, or someone in Tanzania, one of the world's poorest countries, emits in seven months, according to International Energy Agency (IEA) statistics. When Kyoto was agreed, the USA signed and committed to reducing its emissions by seven per cent. But since then it has pulled out of the agreement and between 1990 and 2004 its CO_2 emissions increased by 15.8 per cent. Many states have, however, imposed greenhouse gas regulation schemes. Notably, the Governor of California, Arnold Schwarzenegger, released an Energy Action Plan pledging to reduce reliance on petroleum by boosting efficiency and using more environmental technologies such as solar power and hydrogen fuel cells. The state aims to generate 33 per cent of its energy needs from renewables by 2020, through initiatives such as the "million solar roofs" bill. The State of Massachusetts has introduced targets to reduce CO_2 emissions to 1990 levels by 2010 and achieve a 10 per cent reduction in greenhouse gas emissions by 2020 against 1990 levels.

China

Population: 1296.16 million

2004 CO_2 emissions: 4,732.26 million tonnes of CO_2

Emission reduction target under Kyoto: none, because of developing country status

Changes in emissions 1990–2004: not known

China is currently the world's second-biggest emitter of greenhouse gases, but as a non-Annex I country it is not yet required to cut its emissions. With China accounting for one-fifth of the world's population, its future CO_2 outpourings could dwarf any cuts made by the industrialized nations. *The World Energy Outlook 2006*, produced by the International Energy Agency, predicts that if energy markets evolve without new government measures to alter underlying trends, China will overtake the USA as the world's biggest emitter by 2010. In November 2005, China pledged to provide 15

China's economic boom is rapidly increasing its energy use.

per cent of its energy needs from renewables by 2020, an increase on its previously stated goal of reaching 10 per cent by 2020 and a doubling of today's levels. However, environmentalists say this target is not ambitious enough to offset the climatic damage caused by its rapid economic growth, which will continue to be fuelled predominantly by coal. China is the world's biggest coal producer and its oil consumption has doubled in the last 20 years. Figures produced by the Netherlands Environmental Assessment Agency in 2007 suggest China may already have overtaken the USA as the world's biggest CO_2 emitter. It estimates China produced 6,200 million tonnes of CO_2 in 2006, but said there is still uncertainty about the exact figures.

European Union

Population: 385.92 million

2004 CO_2 emissions: 3,320.47 million tonnes of CO_2

Emission reduction target under Kyoto: –8

Changes in emissions 1990–2004: –2.6

The 15 countries that made up the European Union in 2002 (the EU-15) all ratified the Kyoto Protocol in May of that year. The EU aims to meet its eight per cent target by allocating different rates internally to its member states. Under the scheme, the UK has accepted a 12.5 per cent reduction goal, Luxembourg has the task of reducing its emissions by 28 per cent, and Germany and Denmark have 21 per cent targets. Greece, on the other hand, is permitted a 25 per cent increase, and Portugal a 27 per cent rise. The European Union is the Kyoto Protocol's most enthusiastic supporter, and pressured countries such as Russia, Japan and Canada to ratify the deal so that it could come into force without the commitment of the USA. However, it has struggled to get its own emissions down. Data for the EU-15 (so excluding newcomers Cyprus, the Czech Republic, Estonia, Hungary, Latvia, Lithuania, Malta, Poland, Slovenia and Slovakia) shows 2004 emissions to be –0.6 (without LULUCF) and –2.6 (with LULUCF) of 1990 levels. (LULUCF stands for land-use, land-use change and forestry.)

Russia

Population: 143.85 million

2004 CO$_2$ emissions: 1,528.78 million tonnes of CO$_2$

Emission reduction target under Kyoto: 0 per cent

Changes in emissions 1990–2004: –32 per cent

Of the Annex I countries, Russia was the second-highest emitter in 1990, behind the USA. However, its economy has shrunk considerably since then and, as its industrial activity has fallen, its emissions have dropped by 32

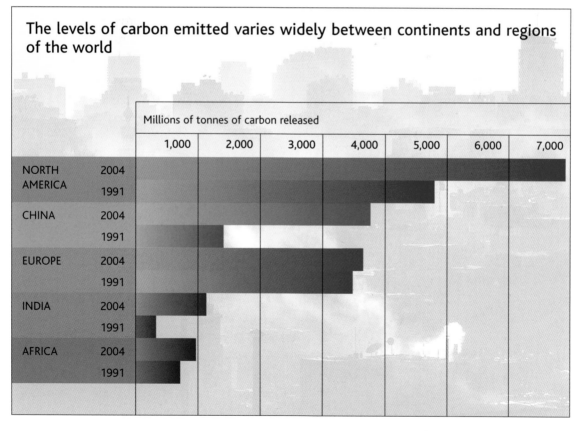

The levels of carbon emitted varies widely between continents and regions of the world

SOURCE: INTERNATIONAL ENERGY ANNUAL 2004

per cent. This means Russia is now well below the level allowed under the Kyoto Protocol, and therefore has the potential to profit by selling its unused emissions entitlement to countries that want to emit more than their permitted concentrations as set by Kyoto.

Japan

Population: 127.69 million

2004 CO_2 emissions: 1,214.99 million tonnes of CO_2

Emission reduction target under Kyoto: –6 per cent

Changes in emissions 1990–2004: +6.5 per cent

Japan is a leading member of Kyoto. In 1990 it was responsible for 8.5 per cent of emissions and its support for the agreement has been critical in the absence of US participation. The use of 1990 as a base year for measuring emission cuts has been disadvantageous to Japan because efficiency-boosting measures put in place following the 1973 oil crisis have left it with little room for added improvement. It committed to reduce emissions by a further six per cent from 1990 levels, but 2004 figures show that its emissions have in fact risen by 6.5 per cent. The country recognizes that its economy could benefit from the Kyoto agreement, as Japanese companies are well placed to capture markets for new, clean technology.

India

Population: 1,079.72 million

2004 CO_2 emissions: 1,102.81 million tonnes of CO_2

Emission reduction target under Kyoto: none, because of developing country status

Changes in emissions 1990–2004: not known

As a developing nation, India is not yet obliged to make any cuts in greenhouse emissions under Kyoto. However, it recognizes that many of its one billion people will be vulnerable to the effects of climate change and therefore ratified the Kyoto Protocol in August 2002. The same year, its prime minister, Atal Behari Vajpayee, rejected calls for poor nations to step up efforts to tackle global warming, saying that countries like India produced only a fraction of the total greenhouse gas emissions, and could not afford to pay the extra costs of cutting them. Half a decade on, with India's economy rapidly developing and its standard of living also rising, its emissions are beginning to climb. They are estimated to have risen by more than 50 per cent in the 1990s and figures for 2004 now rate it as the world's fifth-largest emitter if the European Union's 15 Kyoto members are considered as a single entity.

Low-carbon world

Moving from our current fossil fuel-based economy to a low-carbon one will require a quiver of different technologies and approaches. These include: boosting efficiency when producing energy from fossil fuels; increasing the amount of power generated from non-fossil energy sources; using carbon capture and storage to lock carbon away from the atmosphere; and reducing demand by conserving energy.

Supplying energy is a complex science. It is very expensive to store electricity, so it has to be produced at the moment of demand. If too much or too little is produced, voltage and frequency fluctuations can crash the nation's computers and prompt operators to close down large sections of the network as a precaution as they attempt to maintain a stable system. Yet it is hard to supply exactly what is demanded because our needs vary wildly throughout the day or year. During a major sporting event on television, for example, the demand might be very low throughout the match, and then rocket when everyone gets up to put on the kettle at half-time.

Our network operators have become adept at matching real-time supply to demand, because our current generating mix makes it relatively easy to switch on and off energy supply. However, in a future world powered by many forms of renewable energy, this will be much more of a challenge. Technology and innovation will lie at the heart of the switch to a low-carbon world. At a meeting of energy ministers in London in 2005, the UK Government's chief scientist, Sir David King, told delegates "Energy is the world's biggest industry, and over the next 30 years we'll see the biggest transformation in a century".

A new generation of nuclear reactors may help the UK make the transition to a low-carbon economy.

Boosting efficiency

There are opportunities for improving efficiency at all phases of power generation, transmission and distribution. The World Energy Council's 2006 statement, *Energy Efficiencies: Pipe-dream or Reality*, suggests that, if the planet's existing suite of power plants were replaced with state-of-the-art plants, global CO_2 emissions could be cut by around one billion tonnes a year. Transmission and distribution losses currently average 10 per cent, but in some developing countries losses from illegal connections or non-payment can bring these deficits to 50 per cent. Capturing energy currently wasted could play a huge role in increasing efficiency. For example, the statement suggests that if the natural gas currently flared in Africa were used for power generation, it could produce 50 per cent of the current power consumption of the African continent. A report issued by the McKinsey Global Institute in November 2006 agrees, suggesting that the growth rate of worldwide energy consumption could be halved over the next 15 years given more aggressive energy-efficiency efforts by both industry and households. You can read more about the savings that we as individuals can make in Chapter 5.

Replacing fossil fuels with alternative energy sources

According to the International Energy Agency, in 2004 renewables accounted for 13.1 per cent of the 11,059 million tonnes of oil equivalent of World Total Primary Energy Supply. Of this, combustible renewables and waste (97 per cent is biomass, both commercial and non-commercial) accounted for 79.4 per cent of total renewables, followed by hydro at 16.7 per cent. Only 0.5 per cent of the World Total Primary Energy Supply came from other renewables, such as tidal, wind, solar and geothermal, but these sectors all have the potential to expand in the future.

Nuclear

The memories of accidents such as Chernobyl, concerns over security threats and worries about radioactive waste make nuclear energy a contentious issue. However, in a rapidly warming world, it does offer a means of producing large amounts of power; a two-centimetre nuclear fuel pellet can generate the same amount of power as one and a half tonnes of coal. Nuclear power stations are not carbon neutral, but they do release fewer greenhouse gas emissions than fossil-fuel power stations. The primary source of fuel is uranium, a naturally occurring radioactive element. It works as follows: when a free neutron collides with the nucleus of a uranium atom, the nucleus splits into two smaller atoms and a free neutron. This process is called fission. The free neutron is able to prompt a new fission, and so a chain reaction begins. Because the two new atoms created each time weigh less than the original one, a huge amount of energy is also released. This is harnessed in a reactor as a heat source to turn water into steam. The steam drives a turbine which spins a generator and produces electricity. The amount of energy produced is expressed by Albert Einstein's famous equation $E=mc^2$, where E represents energy, m stands for mass, and c^2 is the conversion factor required to change the mass to energy.

Over the past quarter of a century, the nuclear industry has largely been in decline, but the need to cut greenhouse gases, coupled with worries over the future security of energy supplies, has set several countries rethinking their energy agendas. Finland has begun constructing the first nuclear plant to be built in Western Europe for some 15 years, while China has nine reactors in place and a further 30 commissioned. Sir David King, the UK Government's Chief Scientist, recently called for a further generation of nuclear reactors in the UK to buy time until a new tranche of carbon-neutral technologies are in place.

Scientists have been experimenting with a number of so-called generation IV reactors aimed at making nuclear fission safer and cleaner. There are also hopes that nuclear fusion may be used as an alternative to fission in the future. This process involves combining light nuclei, and has the advantages that light nuclei are easy to come by and the end products of the process are light, stable nuclei rather than heavy radioactive ones. The downside is that, before you can combine light nuclei, you have to overcome the mutual repulsion that exists between positively charged protons in the nuclei. This has so far prevented scientists from creating a viable electricity-producing fusion reactor. However, in 2006 the European Union, along with six other nations, signed a €10 billion (US$12.8 billion) pact to build the world's most advanced nuclear fusion reactor, the International Thermonuclear Experimental Reactor (ITER), in France.

Opposition to nuclear remains high, however. A pole conducted in May 2005 in the UK for the BBC2 programme *Newsnight* found that over half the population – 52 per cent – said it was wrong for the Government to consider nuclear power as a future source of energy. Asked what was the best way to meet energy demands cleanly, 57 per cent suggested renew-able energy while 21 per cent supported nuclear power. In late 2006, protesters were hoping to gather one million signatures against nuclear power in Europe in time for the fiftieth anniversary of the Euratom Treaty, the EU Treaty to promote nuclear energy, in March 2007.

Wind

Wind energy is officially the fastest-growing energy source worldwide, with an average annual growth rate of 23 per cent over the last 15 years, according to the British Wind Energy Association. Its use as a free energy source has a long history. An Arabic document dated 664AD contains the first known reference to a windmill. By the eleventh century people were using windmills extensively in producing food in the Middle East. When the technology spread, via the Dutch, to the New World, settlers relied on the wind to pump water for farms and ranches, and later to generate

National efforts to reduce carbon emissions

The efforts being made by some individual countries to reduce their reliance on hydrocarbons are showing the rest of the world what can be achieved.

ARCTIC OCEAN

1. ICELAND

A leading candidate for becoming a carbon-free economy. One of the most active volcanic regions in the world, Iceland has the Earth's highest concentration of readily available renewable energy sources. These are primarily hydroelectric and geothermal power, which supply over 95 per cent of the country's heat and electricity needs. However, it still imports petroleum to meet 30 per cent of its total energy needs. It hopes to replace the hydrocarbon, which it uses primarily to power road vehicles and its fishing fleet, with hydrogen. Three hydrogen-fuelled buses have been carrying passengers around the streets of Reykjavík since 2003 and there are plans to build a hydrogen production and storage centre and a hydrogen-powered fishing vessel. Iceland's ultimate goal is to become a fully hydrogen-based economy by 2050.

NORTH AMERICA

ATLANTIC OCEAN

PACIFIC OCEAN

SOUTH AMERICA

2. BRAZIL

This vast country derives much of its electricity from hydropower, as well as producing ethanol from sugarcane to fuel its transport. By law, all automotive fuel has to be at least 20 per cent ethanol. As a result, biofuel accounts for at least 40 per cent of all the passenger fuel used in the country and 54 per cent of new cars sold are "flex" cars that can run on pure ethanol or a combination of petrol and ethanol. On the downside, deforestation of the Amazon rainforest remains a major contributor to Brazil's greenhouse gas emissions.

3. CARIBBEAN ISLAND STATES

This region has some of the highest energy costs in the world; as a result many rural communities cannot afford electricity and go without. Grenada, Dominica and St Lucia are among the islands hoping to switch to renewable energy sources. Grenada is examining the possibilities of using wind, hydroelectric, biomass and solar power. Blessed with many rivers and waterfalls, Dominica already generates one-third of its electricity from hydroelectric sources, and aims to meet over 65 per cent of its needs from renewable sources, including hydro, geothermal and wind power, by 2010. Meanwhile, St Lucia hopes to generate at least 20 per cent of its energy from renewable sources by 2010.

4. UNITED KINGDOM

Since 1990, the UK's wind industry has grown to be one of the top 10 in the world. The sector is expanding rapidly and will be the single biggest contributor to the government's aim to supply 10 per cent of its energy needs from renewables by 2010. A record number of wind farms were commissioned in 2006, adding one new offshore and 21 new onshore facilities. As of January 2007, according to the British Wind Energy Association, there were 136 projects operational. That is enough to generate power for over one million homes.

5. DENMARK

Denmark is considered the most advanced user of CHP (combined heat and power) technology in Europe. In normal power plants the heat created while generating energy is lost to the environment, but CHP installations capture this spare heat and distribute it to nearby consumers for warming homes. While typical power plants are usually no more than 50 per cent efficient, CHP units can be 75 per cent efficient, saving energy and greenhouse gas emissions. Across Europe, CHP accounts for about 10 per cent of the electricity market and 10 per cent of the heat market. The European Union aims to increase those numbers to 18 per cent by 2010.

6. SWEDEN

Another contender in the race to be carbon-free, Sweden has pledged to become oil-free by 2020 and hopes to reduce its carbon emissions by 60–80 per cent by 2050. In 2005 it appointed a commission to draw up a plan for reducing its dependence on oil. The commission recommended slashing the use of oil in road transport by 40–50 per cent; cutting that used by industry to 25–40 per cent; and completely withdrawing the use of oil for heating commercial and residential buildings. The country aims to use biofuels, hybrid vehicles, solar cells, wave energy, fuel-cell vehicles and entirely new technologies. It is already pioneering various uses of biofuels and biogas.

EUROPE

ASIA

PACIFIC

OCEAN

AFRICA

7

INDIAN

OCEAN

8. INDIA

India's energy consumption increased rapidly from 4.16 quadrillion British thermal units (quads) in 1980 to 12.8 quads in 2001. This increase is largely the result of the nation's swelling population and speedy urbanization. Currently, coal supplies half India's energy needs, petroleum 34.4 per cent, natural gas 6.5 per cent and hydroelectricity 6.3 per cent. Nuclear and renewables con-tributed a very small 0.2 per cent in 2001. However, the country has a separate government ministry dedicated to non-conventional energy sources, and is look-ing to expand its use of wind and hydropower. It also has one of the largest national programmes for solar energy. With more than 70 per cent of India's population still living in rural areas, har-nessing free energy from the sun is being advocated as a way to bring energy to regions where there is a shortage of electricity. For example, in Uttar Pradesh, where 80 per cent of homes have no power, a project is under way to power 2,000 computers in primary schools across 70 districts. In other areas solar energy is used for cooking, heating water, light and running tube wells. In Bangalore, some apartments, hostels, and hotels are beginning to use a solar-powered steam-based cooking system to provide food for up to 500 people at a time.

7. KENYA

Kenya began exploiting its geothermal resources in the 1980s and now hosts Africa's largest geothermal plant at Ol Karia. The government had previously favoured hydroelectric schemes for generating electricity, but the geothermal system proved more reliable when drought struck in 2001 and hydroelectric systems failed as reservoirs behind dams dried up. The year after the drought, work began on a second geothermal plant at Ol Karia; a third will be up and running by 2010.

SOUTHERN

OCEAN

ANTARCTICA

The lattice design of some early wind turbines caused the death of birds that used them as perches.

electricity. Steam power later overtook the wind as a prime energy source, but during the fuel crisis of the 1970s wind turbine generators re-emerged as a viable means of producing electricity. Today, climate change has firmly put this non-polluting, renewable energy source back in the spotlight.

As a result, the industry is expanding apace around the world. In the past decade, the global capacity rose from 2,500 megawatts (MW) in 1992 to almost 75,000MW at the end of 2006. Almost three-quarters of this increase is due to new installations in Europe. In Denmark, wind energy now supplies 20 per cent of its energy needs; in Spain and Germany it contributes five per cent. Regionally, these figures can be higher. In the north German state of Schleswig-Holstein, wind power meets 30 per cent of the region's total electricity demand, while in Navarra, Spain, half of consumption is met by wind power. The European Wind Energy Association predicts that there will be enough wind farms up and running by 2020 to supply more than 12 per cent of the EU-15's energy needs. That is enough to satiate the needs of 195 million household consumers. Outside Europe, the technology is also on the rise. In August 2006, Cape Town signed a deal for South Africa's first ever commercial wind development. In the same month, the American Wind Energy Association announced that the USA's wind capacity exceeded 10,000MW in 2006, enough to power 2.5 million homes.

Most wind farms today use turbines comprising a single tapered pole with a three-pronged rotating blade attached to the top. The wind turns the blades at between 10 and 30 revolutions per minute. This then spins a shaft, which connects to a generator and makes electricity. The height of the pole, width of the blades and arrangement of multiple turbines vary according to local conditions. The largest turbines stand over 100 metres high, and have blades spanning 80 metres. In December 2006, the UK Government gave the green light to plans for the world's largest offshore wind farm. The London Array will consist of 341 turbines over 233 square kilometres between Margate in Kent and Clacton in Essex. The developers claim that the wind farm will be powerful enough to supply one-tenth of the Government's target of 10 per cent from renewable energy sources by 2010. It will avoid 1.9 million tonnes of carbon entering the atmosphere each year.

There has been some opposition to wind farms, primarily from residents who consider developments in remote locations could reduce the natural beauty of the landscape, and environmentalists who fear the impacts on wildlife and habitats. More than a decade back, in the early days of renewable energy, two wind farms at Tarifa, Spain, and Altamont Pass, California, were found to be killing hundreds of birds each year. The former was located on a major migratory route, and the latter within a mountain pass inhabited by golden eagles and other raptors. Birds found the lattice design of the turbine towers perfect as perches and nesting sites, and frequently flew up from the towers into the blades. Today, more streamlined designs and tougher planning regulations demanding detailed environmental impact assessments have reduced the risks to birds. A report conducted by the Centre for Evidence-Based Conservation at the University of Birmingham, UK, found that wind farms can have a negative effect on relatively large and slow-moving ducks and waders, but that other birds were not significantly affected.

There are considerable efficiency benefits to be gained from making individual turbines as large as possible, because the wind flows faster and with less turbulence at greater heights above the ground. This, coupled with the fact that modern wind farms are often extensive operations, has increased opposition to new installations. A scheme planned for the Isle of Lewis, Scotland, demonstrates the pros and cons that such developments can bring. The developers initially wanted to install 234 turbines, each 140 metres high on an internationally important and protected wetland. For certain species, including Golden Plover and Dunlin, the peatland habitat is the best in Europe, hosting a significant proportion of their total British or world populations. Following complaints, including vociferous objections from the Royal Society for the Protection of Birds, the company planning the farm reduced the number of turbines to 181. Local residents have been in two minds about the scheme. On the one hand they feel the wind farm could reduce tourism by blighting a renowned area of outstanding natural beauty. But on the other, the traditional industries of the Western Isles are in decline and the wind farm will provide jobs and the opportunity for a remote part of Scotland to reinvent itself as a leading force in the expanding renewable energy market.

A driver fills up with biofuel. Biofuels are made from crops such as wheat, corn and sugar cane.

Biomass

The term biomass is used to describe all recently living plant and animal material that can be used to produce energy or fuels. Biomass can produce electricity, heat, liquid fuels, gaseous fuels, and a variety of useful chemicals, including those currently manufactured from fossil fuels. Fossil hydrocarbons such as oil, gas and coal, which have formed through geological processes, are not considered to be biomass, despite being generated from plant material. Both hydrocarbons and biomass release CO_2 when processed, but biomass energy sources are considered carbon-neutral. This is because they take carbon out of the atmosphere while living and release it back when burned as part of a carbon cycle taking place over a matter of months or years. Because plants that ended up as fossil fuels took carbon out of the atmosphere millions of years ago, when they release their carbon to the atmosphere it is in addition to that being currently recycled through the plants, soils, oceans and atmosphere. This is how the concentration of CO_2 in the atmosphere has risen inexorably since industrial times.

Two advantages of using biomass are that it is the only sustainable energy source that does not operate intermittently, and it can potentially generate energy for heat, power and transport from the same installation. However, it differs from fossil-fuel energy sources in that the diversity of primary materials calls for a range of processing technologies. Waste biomass materials suitable for producing energy include wood chips and sawdust from forestry residues, farm slurries, poultry litter, fisheries and slaughterhouse wastes, plus organic municipal wastes. Then there are various crops with the potential to yield high energy levels if harvested every few years, a process called short-rotation coppicing. The woody grass *Miscanthus*, willow and poplar are all examples; 22,000 tonnes of *Miscanthus* can provide enough electricity to power 2,000 homes. The methods used to treat the varying starting materials include simple combustion, co-firing, pyrolysis (heating to decompose the fuel), gasification (heating in a flow of gas, usually air or steam), anaerobic digestion, and fermentation.

Worldwide, biomass is the fourth-largest energy resource after coal, oil and natural gas. In 2004, combustible renewables and waste supplied 10.6 per cent of the world's energy, slightly less than in the 1970s. In developing countries, biomass contributes a higher proportion of the energy supply, as firewood and dung are still heavily relied on in rural areas beyond the reach of electricity networks. There is potential for biomass to become a more significant contributor of renewable power in the developed world. In Europe, biomass accounts for four per cent of energy used, which is around two-thirds of the renewable market. The main objective of a Biomass Action Plan, put forward in December 2005, is to double this share by 2010. In doing so, the EU hopes to reduce oil imports by eight per cent, prevent greenhouse gas emissions worth 209 million tonnes of CO_2 equivalent per year, and create up to 300,000 new jobs in the agricultural and forestry sector. In the UK, the Biomass Task Force that reported to the Government in 2005 suggested that a million hectares of land may be available for non-food uses, which, on current yields, could provide around eight million tonnes of energy crops.

Biomass crops are also a useful source of fuel. First developed following the oil crisis of the 1970s, biofuels are slowly becoming a global commodity. They are derived from crops such as wheat, corn, sugar cane and oilseed rape. The two main types are ethanol, used in petrol engines, and vegetable oil methyl esters (VOME), used in diesel engines. Of the two, ethanol is the most prevalent. Ethanol is used pure, in a blend or in its ether form. If it is used pure or in a very high concentration, for example at 85 per cent in the fuel E85, then traditional petrol engines have to be modified. A number of car manufacturers including GM, Ford, Daimler-Chrysler, PSA and Renault now sell vehicles in Brazil and the USA that run on E85. Unmodified engines can run on blends in which just five or 10 per cent of ethanol is added to the fuel. VOME fuels are made from vegetable oils, such as rapeseed, sunflower, soybean or palm. Like ethanol, they can be used pure or blended. Vehicles have to be modified if the fuels are used pure; mostly they are incorporated in blends in concentrations of up to 30 per cent.

Brazil and the USA accounted for most of the global production of biofuels in 2003. In 2005, Brazil's car-makers sold more vehicles adapted to run on alcohol than conventional petrol-driven models. "Flex-fuel" cars, which run on any combination of ethanol and petrol, accounted for 54 per cent of the Brazilian market. A biofuels pioneer, Brazil began a Pro-Alcohol

Solar energy, such as that generated via rooftop panels, currently accounts for just 0.03 per cent of the world's energy.

programme more than 20 years ago to encourage the use of sugar cane-derived ethanol as a vehicle fuel, to reduce its dependence on Middle Eastern oil. Europe has historically lagged behind in the race to develop biofuels and is now playing catch-up. Since 2000, Germany has begun developing a market in VOME, while Spain and Poland have been experimenting with ethanol. In 2006, the UK announced that its first bioethanol plant would soon be annually converting 340,000 tonnes of locally grown wheat into 131 million litres of ethanol. If the Government's Renewable Transport Fuel Obligation is to be met, which demands that five per cent of all motorcar fuel must come from renewable sources by 2010, a further nine such plants will be needed.

In recent years, Sweden has also been expanding its use of biofuels in line with its aim to become oil-free by 2020. It runs a state-backed bioethanol programme, by which drivers pay no duty on the fuel. In addition, E85-enabled cars may park free of charge in Gothenburg, Stockholm and other municipalities. Biofuel cars are 20 per cent cheaper to insure and are exempt from the Stockholm congestion charge, while both personal and fleet users pay less tax. While this is costly for the Government, it has helped it to generate jobs in rural areas, develop expertise in an emerging industry and reduce its reliance on fossil fuels. The country's extensive forests provide a vast potential biomass source. Göteborg Energy is presently building a biomass gasification plant, which will convert woodchips to methane in sufficient quantities to power 75,000 cars. Sweden has also pioneered some other more unusual biomass sources; the train that carries passengers between Linköping and Västervik runs on biogas (methane) produced from the entrails of dead cows. One animal generates enough gas to keep the train running for four kilometres.

Solar

Despite the ubiquitous nature of sunshine, solar energy production currently accounts for just 0.03 per cent of the world's energy needs. A basic problem is that solar power has a low conversion efficiency. However, demand has grown at about 25 per cent per annum over the past 15 years. During 2005, global production of solar batteries rose by 47 per cent by volume, and this pace is believed to have stepped up again in 2006. Japan, the USA and Germany are the market leaders; around half of the world's solar cell production was manufactured in Japan in 2003, while the USA accounted for 12 per cent. Experts have predicted that 2018 will be the turning point for solar power, when production costs are forecast to fall far enough for the sector to be competitive with other energy prices without subsidies. However, there is currently a shortage of poly-crystalline silicon, the basic material that is used to convert the sun's rays into electricity.

There are three main ways in which the sun's energy is harnessed. Photovoltaic cells are used to convert light directly to electricity. These are the kind of cells that power our pocket calculators. If enough cells are combined in an array, they can power cookers, buildings or even orbiting satellites. A second use of solar energy is for directly heating water. Thirdly, solar furnaces use arrays of mirrors to focus the sun's rays in one place and produce very high temperatures. This technology works on the same principle as when you use a magnifying glass to start a fire. Australia is currently planning to build one of the largest "solar concentrators" to power 45,000 homes. Comprising hundreds of square metres of curved mirrors that will concentrate heat onto photovoltaic panels, the power station is likely to generate temperatures of 1,000°C. Similar large-scale projects are already under way in the USA and Portugal. Two German scientists, Dr Gerhard Knies and Dr Franz Trieb, believe that covering just 0.5 per cent of the world's hot deserts with such schemes would provide the world's entire electricity needs.

Hydropower

Hydropower provides 2.2 per cent of the world's primary energy supply. It has long been used for simple mechanical tasks such as grinding corn; the Domesday Book, a survey of life in England in 1086, lists 5,642 waterwheel-driven mills in England south of the River Trent. Hydroelectric power supplied 16.1 per cent of global electricity in 2004, down from 18.5 per cent in 1994. According to the Worldwatch Institute, the sector's output rose five per cent in 2004 after a period of stagnation. Although large-scale hydropower remains one of the least expensive ways to generate

The Three Gorges Dam on the Yangtze River is the largest hydropower project in the world.

electricity, it is not a very environmentally friendly option. Building dams often involves moving communities and flooding their former settlements, along with disrupting natural river flows and ecosystems. The lack of suitable sites has hampered further development of large-scale dams in many countries.

Most hydropower projects have four components. A dam built across a river provides a controllable flow of falling water. The force of falling water pushes against a turbine, turning the blades and converting kinetic energy into mechanical energy. The turbine connects with a generator which also spins, and in doing so turns mechanical energy into electrical energy. Finally, transmission lines feed this remotely produced power into homes and businesses. Hydropower supplies at least 50 per cent of electricity production in 66 countries, and at least 90 per cent in 24 countries, according to Australia's Research Institute for Sustainable Energy. Its potential is likely to be exploited further in the hunt for renewable resources, primarily in developing countries. China is leading the way. It is home to the largest and most ambitious hydro project in the world, the Three Gorges Dam scheme, which involves harnessing the Yangtze River, Asia's longest, to generate almost a tenth of the nation's electricity needs.

Due for completion in 2009, the project has drawn intense opposition. It has forced some 1.9 million people from their homes and has the potential to cause an ecological disaster by silting, and thus depriving agricultural land and fish downstream of nutrients. However, it will provide an amount of power equal to several coal-fired power stations.

Wave and tidal power

People first built tidal mills in the Middle Ages, trapping incoming flows of water in storage ponds and using the energy of the ebbing water to turn mill wheels. However, the expense of developing technologies for large-scale sea-based operations has prevented the widespread adoption of tidal power. There are currently only a handful of such plants operating in the world. The largest of these, which has a capacity of 240 megawatts, is located in the estuary of La Rance, near St Malo, in France. Another facility, in the Bay of Fundy, Nova Scotia, Canada, draws on the highest tidal range in the world to power 4,000 homes. The UK's Severn River boasts the

Iceland's Svartsengi power plant harnesses geothermal energy.

second-highest tidal range but has yet to be exploited. Every 12 hours, the incoming tide whips around southern Ireland and is then squeezed between South Wales and the coasts of North Devon and Cornwall. As the channel narrows, the water is rapidly funnelled towards the mouth of the River Wye, where the tidal rise reaches 15 metres. Developers believe a barrage built across the estuary could provide as much as seven per cent of the power needs of England and Wales. In light of climate change stepping up the pressure to find new sources of power, the Government recommended a further study of the £14 billion project in 2006, but said it would raise "strong environmental concerns".

Aside from barrage schemes, companies around the world have been experimenting with extracting power from waves and tidal streams. Wave-based devices generate electricity from movements of the sea surface, whereas tidal stream installations sit on the sea floor and use the regular ebb and flow of tides. The first commercial subsea power station to harness the tidal currents began supplying energy for the town of Hammerfest, Norway, in 2003. The device uses the tidal energy of the sea in the same way that wind turbines tap into the power of air currents. The generator has 10-metre-wide blades that are fixed to a 20-metre steel column attached to the seabed. They rotate as the tide flows over them, driving a generator to produce electricity. The power station currently provides enough energy to power 30 homes, but there are hopes that a further 20 stations could be developed nearby. A report published in 2006 suggested that the UK could generate one-fifth of its energy needs from such technologies. Only a handful of wave and tidal stream devices are currently installed. Scotland has two prototype wave machines, the Pelamis deepwater system and the Limpet shoreline device, while Marine Current Turbines has been testing prototype tidal stream generators in Devon and Northern Ireland. The company plans to have a commercial product ready by 2007 or 2008.

Geothermal energy
Geothermal energy is literally the heat contained within the Earth that drives geological processes on a planetary scale. We see evidence of this heat at the surface in the form of bubbling hot springs, volcanoes and geysers. If you travel towards the Earth's core, the temperature rises by about 2.5°C or 3°C for every 100 metres. The first attempt to generate power from geothermal energy took place in Larderello, Italy. Its success soon prompted Japan, the USA, New Zealand and Mexico to follow suit. Today, many countries harness the heat emanating from the centre of the Earth, usually by tapping into underground water supplies held as steam under high pressure. Although geothermal power plants supply only 9.3 gigawatts (GW) of capacity worldwide, against the total renewable power generation of 4,100GW, supply is forecast to grow by 10 per cent a year up to 2010. According to the International Geothermal Association, this is more than three times the rate of the last 10 years. The USA is currently the largest exploiter of geothermal energy, but the Philippines is catching up fast.

Hydrogen

Geothermal energy can be used to help generate power from hydrogen. Hydrogen gas is lighter than air and therefore exists on Earth only within compounds such as water (H_2O) and methane (CH_4). It can be made from various sources including water, fossil fuels and biomass. There are currently

Some scientists have suggested sprinkling the Southern Ocean with iron, to aid the growth of CO_2-absorbing plankton.

two methods used to isolate hydrogen. Steam reforming separates the hydrogen atoms from carbon in methane, but this process results in some greenhouse gas emissions. Alternatively, electrolysis splits hydrogen atoms from water. This technique only releases steam, but is presently very expensive. The hydrogen is passed through a fuel cell that powers an electric motor. Reasonably small fuel cells can power electric cars. Blessed with copious resources of geothermal and hydrothermal power, Iceland is leading the world in developing hydrogen technology without the use of fossil fuels. The country has been experimenting with using hydrogen to power buses, and in 2003 opened the world's first commercial hydrogen filling station, in Reykjavik. Ultimately, it hopes to give up its reliance on fossil fuels and become a hydrogen-based economy.

Capturing carbon and mega-engineering

As well as investigating the opportunities that renewables offer to cut carbon emissions, governments are researching ways to capture carbon and store it away from the atmosphere. One option is to apprehend the CO_2 produced by power stations by passing it through "amine scrubbers" that react with and trap the gas. Another option is to clean up the fuel before it is burned. For example, it is possible to remove pollutants and then turn coal into hydrogen using a process called gasification. The gas is burned in a turbine to generate electricity. The heat from this initial combustion is then used to raise steam for a second turbine stage, thus providing two bites at the thermal cherry. This method of using coal is much more efficient than current processes.

Captured CO_2 can be compressed, then injected under pressure down pipelines into disused coal seams, old oil or gas wells or porous rocks filled with saltwater. The idea is to lock the gas away deep underground or in subsea stores, so that it never has a chance to reach the atmosphere. Oil company Statoil already removes millions of tonnes of CO_2 from natural gas at the Sleipner gas field in the North Sea and buries it in a saline aquifer without bringing it onto dry land. The IPCC published a report on the state-of-the-art of carbon capture and storage technologies in 2005. It suggested that, by 2050, "around 20–40 per cent of global fossil fuel CO_2 emissions could be technically suitable for capture, including 30–60 per cent of CO_2 emissions from electricity generation and 30–40 per cent of those from industry".

Other suggestions for actively cooling the Earth involve "mega-engineering" projects. These include: reflecting more of the sun's rays by placing a 1,000-kilometre-wide mirror between the Earth and sun, or sending floating, reflective, aluminium balloons into the stratosphere; using aeroplanes to spray sulphate particles into the atmosphere to seed clouds; placing spray turbines across the oceans to cast salt spray into the air and whiten clouds; and seeding the Southern Ocean with iron or urea to encourage more CO_2-absorbing plankton to grow.

CHAPTER FIVE

Taking responsibility for our actions

HOW INDIVIDUALS CAN MAKE A DIFFERENCE

The demand for energy in expanding cities such as Shanghai, China, is rising fast.

Around half the world's six billion people live in cities. As the global population continues to climb, and those in developing countries strive for the higher standards of living enjoyed by their Western cousins, this is expected to rise to 60 per cent by 2030. The demand for energy will rise concurrently, because urban areas suck up much greater quantities of power than rural ones. In China and India, the world's two most populous nations, each with a billion-plus citizens, the increase is expected to be rapid. Their growing middle classes are already swapping bicycles for cars, moving into larger, air-conditioned houses and becoming eager consumers of electrical items such as fridges, computers, DVD players and stereos.

Not surprisingly, citizens of these countries of the developing world are hungry for a share of the wealth that Western nations have enjoyed for decades.

As our appetite for energy strengthens, the planet's ability to satiate it is weakening. The preceding chapters have shown that industrialization has pumped enough greenhouse gases into the atmosphere to potentially change our planet irretrievably within years. We once considered the world's varying climates and geographical environments to be constant and predictable, but today that comforting certainty is slipping through our fingers. The predictions made by the world's top climate scientists of higher sea levels, melting ice caps and extreme weather are beginning to become reality. And their forecasts for runaway climate change and catastrophic changes to the planet no longer sound like alarmist outcries but rather grave warnings worth heeding. It is now clear that only a concerted and immediate effort, at international, national, regional and individual levels, will give us any chance of keeping the ailing world's temperature beneath a threshold that could flip its natural systems into new, uncertain states.

Embracing change

The move to a non-carbon-based global economy will prompt many changes to the way we live. The cities that will become the dwelling places for most of the world's people in future will be very different from today in their use of energy. Distant power stations, that lose 60 or 70 units of energy for every 100 put in, will be a thing of the past. Instead, our heating and lighting will come from locally installed combined heat and power (CHP) units, vast offshore wind farms, and a new generation of nuclear power stations, perhaps backed up by personal wind turbines and solar panels. Whereas today, no one really knows how much energy they use on a day-to-day basis, in future we will monitor our use much more closely. For inhabitants participating in Chicago's Community Energy Cooperative, this is already a reality. Most people do not realize that the price of electricity fluctuates on a daily basis as demand rises and falls. But the cooperative members get an email warning them when their energy is likely to be pricey. Just as mobile phone users delay calls until the weekend to avoid peak-time prices, so these participants simply turn down the air-conditioning or postpone running the dishwasher for a few hours. Not only does the scheme help them save money on their utility bills, it promotes a culture of energy conservation.

New buildings will be designed to make use of the sun's energy and will be insulated so that wastage is kept to a minimum. There are several developments already in place that demonstrate how zero-carbon housing can be achieved.

The Hammarby Sjöstad development in Stockholm is an example. The suburb, whose name means "the town around the lake of Hammarby Sjö"

CALCULATING YOUR CARBON FOOTPRINT

Experts estimate that, to stop climate change, each person on Earth should cause a maximum of 2.5 tonnes of CO_2 emissions per year as a total of every aspect of their lives. In the UK this means cutting carbon use by 80 per cent. There are several websites that allow you to assess accurately the amount of carbon you are responsible for. Try SafeClimate's carbon calculator at http://safeclimate.net/calculator/. A project of the World Resources Institute, SafeClimate quotes greenhouse gas units of carbon dioxide, as well as converting other greenhouse gases into units of CO_2 based on their relative global warming potentials. Alternative carbon calculators include:
- http://www.carboncalculator.com/
- http://www.carboncalculator.co.uk/

The Guardian published a good guide to assessing your carbon footprint on its website:
http://environment.guardian.co.uk/ethicalliving/story/0,,1997277,00.html

HOW MUCH CO_2 OUR ACTIVITIES CAN GENERATE
(per person per year unless otherwise stated)

- Use of TV, Videos, stereos: 35kg
- Heating: every degree on the thermostat accounts for 25kg of CO_2 each year
- A meal at a restaurant generates 8kg per diner
- Washing clothes: 100kg; drying clothes in a tumble dryer: 36kg
- Taking a bath: 50kg
- A short break to Prague: 220kg
- Trips to the seaside: 200kg
- Charging a mobile phone: 35–70kg
- Sending a letter: 0.01kg

has been converted from an old industrial and harbour area into a modern neighbourhood built on sustainable principles. Once the area is fully developed, in 2015, it will have 10,000 residential homes for just over 20,000 people, and a total of 30,000 people will live and work there. Today, over 10,000 people live there, mostly on the south side. The neighbourhood comprises collections of blocks, divided by wide, tree-lined avenues. Some of the balconied apartments are coloured externally in cream, terracotta and pale blue, others have walls coated in sleek, reflective photovoltaic cells. Trams glide between the low-rise blocks, people cycle on designated paths, and families meet for lunch at waterfront cafes.

Hammarby has its own wastewater plant, where waste is purified and recycled into fertilizer for agricultural land. Its energy comes from a thermal power plant that supplies district heating and cooling from treated waste-water and biofuels. Some biogas is generated from food waste, and biogas cookers are installed in roughly one thousand apartments. A transport system has been developed with the aim of reducing car use; the Lotten ferry carries passengers across the lake every ten minutes, and there are frequent buses and trams to the city centre. A car pool system, "City Car", is open to all residents and workers and currently has around 350 members and 25 cars. Waste is collected using an underground collection system, where items of rubbish are sucked through pipes into a central room, one fraction at a time. Excess rainwater is collected, purified using sand filters and artificially created wetlands, then released into the lake of Hammarby Sjö.

A similar ideology lies behind the Beddington Zero-Energy Development (BedZED) in South London. This is the UK's biggest eco-village, comprising a high-density development of 99 residential properties ranging from one-bedroom flats to four-bedroom houses, plus 1,450 square metres of office space. The primary aims in building BedZED were to reduce the amount of energy required by the housing and office units, and then to supply the necessary power through on-site renewables.

The office space is located on the north side of the blocks and the residential units on the south side, because you generally need to heat houses more than you need to cool them, and cool offices more than you need to heat them. Each terraced block is built with a south-facing triple-storey glass "sun-space". The idea is that the sunlight enters these spaces throughout the year, even at winter solstice when the sun is at its lowest point in the sky. Because the internal concrete walls have a high thermal mass, heat coming in from the sun is stored in the building fabric and re-released throughout the day. Each block is also wrapped in a thermal jacket of 300-millimetre rock wool insulation. The outcome is that the internal temperature remains at about 18–21°C throughout the year, without the need for central heating.

A ventilation system keeps fresh air circulating through the houses. The roofs are topped by a series of multicoloured wind cowls, which look a little like the combs of giant metallic cockerels. As the wind blows, fins catch the oncoming wind and swing round, drawing cool, fresh air into the houses through air vents positioned low-down on room walls. At the same time, warm, stale air rises and is expelled through ducts positioned higher up on the internal walls. When there is little or no wind, the ventilation works on a "passive stack" system, with the rising warm air naturally drawing in cooler air. There is a heat exchanger within the cowl so that, as the cool air comes in, it passes over the outgoing warmer air and is slightly pre-heated.

All units have light fittings that can only be used with low-energy compact fluorescent light bulbs, and all the appliances are A-rated for energy efficiency. This has reduced the electricity needs of BedZED inhabitants by 33 per cent. Electricity is usually provided by an on-site combined heat and power unit located in a separate building, plus 777 square metres of photovoltaic cells positioned on the roofs and windows of the terraces. Other features aimed at reducing environmental impacts include using aerated water in taps and low-flow shower heads to reduce water use; planting sedum, a kind of succulent that grows like a living carpet, on roofs to reduce run-off and increase biodiversity; and gathering rainwater in storage tanks for use in the low-flush toilets. Kitchen units are made from beech wood, so contain no formaldehyde, and all the paints used in the initial decoration of BedZED were eco-friendly. Split waste bins in the kitchens make recycling easy; BedZED aimed to cut waste by 60 per cent.

A report published in 2006 by WWF concluded that the average person in the UK has an ecological footprint of 5.4 global hectares. A global hectare is the equivalent to a normal hectare but adjusted for average global productivity. The figure signifies that we each need that amount of land to produce the food, clothing, energy, water and other materials we consume throughout our lives. The report concluded that, if the world's six and a half billion inhabitants all began consuming at the same level as the UK's inhabitants, we would need three planets to support us. BedZED embraces the ideology of "one planet living", a ten-point formula that will inform future eco-developments in Australia, China, North America and South Africa. The 10 points to ensure we can all live comfortably on just one planet, rather than three, are: zero carbon; zero waste; sustainable transport; use of local and sustainable materials; use of local and sustainable foods; sustainable water supply; natural habitats and wildlife; culture and heritage; equality and fair trade; and health and happiness.

Ultimately, it is likely that we shall all have some form of personal carbon allocation. Use anything above our allowance and we will have to pay extra; be frugal and we will be able to sell any excess. In 2005, the Governor of the State of Oregon, USA, set up a task force to design a carbon allocation programme for the state, and the UK is also considering adopting a rationing scheme. Under the UK scheme, first mooted in July 2006, all citizens would be allocated an identical annual carbon allowance, stored as points on an electronic "swipe" card similar to "air miles" or supermarket loyalty cards. Points would be deducted at point of sale for purchases of non-renewable energy such as petrol and airline tickets, as well as food. People not using their full allocation, such as families not owning a car, would be able to sell their surplus carbon points into a central bank. The results of a feasibility study that investigated how the scheme might work were published in December 2006. It proposed a hybrid system using permits and taxes, with the permits possibly issued, tracked and traded through the existing banking system using chip and pin technology. It could be working within five years.

In the decadent West, the idea of having limits imposed on our wasteful, energy-consuming lifestyles will probably come as a shock to many people. But the rapid rate and scale of climate change are such that radical changes are required. Presently, 33 per cent of people have 94 per cent of the global dollar income and account for 90 per cent of the global historical total of greenhouse gas emissions, whereas the other 66 per cent have six per cent of global dollar income and are responsible for only 10 per cent of emissions. Our present global economy, with its cornerstones of hydrocarbons and consumerism, seems entirely incompatible with aims to balance inequality, preserve dwindling resources, conserve biodiversity and, above all, halt dangerous climate change.

Individual responsibility

Leisure activities, including watching floodlit sporting events, account for a fifth of an average person's carbon emissions.

While the world debates how best to shift to a low-carbon economy, we can start to make a difference by trimming the extent of our own "carbon footprint". This is the amount of carbon we are responsible for putting into the atmosphere as we go about our daily lives. Direct inputs come from the hydrocarbons burned when we drive our cars, take flights abroad, and use electricity and heating in our homes and offices. Indirect uses relate to the products we consume. Almost anything we buy – from food and clothing to electrical equipment – will have gobbled up fossil fuels during its manufacture and distribution. The more we consume, and the more the items we buy are grown or produced far from our homes, the bigger our carbon footprint. It is a good proxy for wealth. The carbon footprint of the average US citizen is around 19 tonnes of CO_2, while the equivalent British citizen is responsible for 10.92 tonnes.

But the carbon footprints of some developing countries are much lower. The world average is around six tonnes per person.

A study by the Carbon Trust, published in 2006, looked in depth at how different elements of the average British person's lifestyle contributed to their overall carbon footprint. It found that:

- 1.95 tonnes, one-fifth of our annual CO_2 emissions, are generated through leisure activities. These range from taking a holiday by car, visiting a gym or heated swimming pool, or watching an evening football match under floodlights. This category also includes the use of TVs, DVD players and stereos.
- 1.49 tonnes come from heating our homes.
- 1.39 tonnes are generated by food and catering, ranging from direct emissions through freezing, refrigerating and cooking food, to indirect emissions created during the production of food and drink items and services. Production was considered to include cultivating raw ingredients, making and using packaging materials, distributing food products, plus disposing of, or recycling, packaging.
- 1.37 tonnes are generated through household activities, including lighting, using household appliances such as vacuum cleaners and DIY equipment, plus the electricity used to create the building and its interior furnishings.
- 1 tonne comes from buying, washing and drying clothing and shoes.
- 0.81 tonnes are emitted through commuting.
- 0.68 tonnes are generated through the flights we take.
- 0.49 tonnes come from education, including the production of books and newspapers.
- 0.1 tonnes are from CO_2 emanating from equipment such as mobile phones and computers.

Reducing your carbon footprint

Within the home

The first thing to work out is how running your home contributes to your carbon footprint. For homes in the UK, the Energy Saving Trust offers a home energy check via an online questionnaire (http://www.est.org.uk/). It uses your answers on the types and age of the house, variety of heating system installed, fuel used and present insulation, to show how you can make financial as well as carbon savings.

For a more comprehensive report, you can pay to have a green audit of your home. Costing around £160 for a three-bedroom house, a "green expert" will assess your current carbon footprint, suggest where improvements can be made, and offer contacts to help achieve the reductions. Examples of companies offering this service in the UK are The Carbon Co$_2$ach (http://carboncoach.com/) and 3 Acorns Environmental

Reducing water wastage and keeping your water cylinder thermostat below 60°C can help reduce your carbon footprint.

Transformations (http://www.3acorns.co.uk/). The latter is run by Donnachadh McCarthy, who is author of *Saving the planet without costing the Earth – 500 simple steps to greening your lifestyle* (published by Vision Paperbacks). For an in-depth guide to going "low-carbon", read *How to live a low-carbon life: the individual's guide to stopping climate change* by Chris Goodall (published by Earthscan). Another useful website with tips on how to make all aspects of your home more energy-friendly is http://www.whatyoucando.co.uk/.

Low-cost changes
There are several simple actions you can take to reduce your home's carbon emissions without much effort or investment required. These include:

• Turning the thermostat down by 1°C to cut heating bills by 10 per cent. Doing this could cut your annual carbon footprint by 400 kilograms. Ideally the thermostat should be set at between 18°C and 21°C.
• Setting your water cylinder thermostat to no higher than 60°C.
• Closing curtains at dusk to stop heat escaping through the windows.
• Turning lights off when you leave a room.
• Never leaving electrical items on standby or unnecessarily charging.
• Using the half-load functions of dishwashers, washing machines and tumble dryers if you have to wash a small load; washing clothes at 30°C; avoiding using the tumble dryer altogether by drying clothes outside.
• Boiling only as much water as you need, in an insulated kettle.
• Fitting low-energy light bulbs.
• Having a shower instead of a bath.
• Switching to a green energy provider. Many suppliers offer some form of green tariff but they have varying credentials. In the UK, Good Energy Ltd

is the only company that supplies 100 per cent of its energy from renewables. Ecotricity was the UK's first renewable supplier. If you sign up for its "new" tariff you will be helping it to invest in new renewable technologies, but only 25 per cent of the electricity it supplies comes from renewable sources. You will generally pay extra to obtain your electricity through renewable means. npower Juice charges standard rates and feeds one unit of electricity into the national grid for each one used by Juice customers, primarily from the North Hoyle Offshore Wind Farm. You can find out more about other green suppliers at: http://www.greenelectricity.org/tariffs.php Similar deals are on offer elsewhere. For tariffs available in the USA, see http://www.eere.energy.gov/greenpower/; in Canada, see http://cleanairrenewableenergycoalition.com/; in Switzerland, see http://www.naturemade.org/; in Australia, see http://www.greenpower.gov.au/home.aspx; and in Belgium, Netherlands, Sweden or Germany see http://www.greenprices.com/eu/index.asp.

Higher cost changes

Spending money now on the following measures could help you make savings in the long run:

• Replace your existing boiler with a condensing boiler. From April 2005, almost all new boilers installed in England and Wales must be condensing boilers in the A or B efficiency bands. This type of boiler can save up to one-third of typical annual heating bills. According to the Energy Saving Trust, if everyone in the UK with gas central heating installed a high-efficiency boiler, we would save enough energy to heat 3.7 million homes for a year.

• Choose electrical appliances with the highest possible energy rating. The EU energy label rates products from A (most efficient) to G (least efficient), apart from refrigerating products which are rated up to A++. By law, the label must be shown on fridges and freezers, washing machines, dishwashers, electric ovens and light-bulb packaging. In the UK, look out for the Energy Saving Recommended logo. Buying an energy-efficient fridge-freezer to replace an inefficient model could cut carbon dioxide emissions produced indirectly by your home by up to 190 kilograms a year.

• Insulate your home. Uninsulated walls account for one-third of the heat lost from homes. Filling cavity walls can save up to a tonne of CO_2 emissions a year. There are grants available in the UK for people wishing to insulate cavity walls. Find out more from the Energy Saving Trust at www.est.org.uk.

• Build a south-facing conservatory to make the most of solar power – but don't heat it.

• Generate your own renewable energy by installing a wind turbine or solar panels. There are two options for solar power: water-heating collectors, which absorb energy from the sun and transfer it to heat water, and photovoltaic or solar electric panels, which transform the solar radiation directly into electricity. For maximum efficiency, solar panels should be mounted on a south-facing roof at a 30° angle to the horizontal, and away from shadows cast by trees, surrounding buildings or chimneys. Find out more from the Solar Trade Association at http://www.green energy.org.uk/sta/. It may be possible to sell back to the national grid any energy that you generate in excess of your needs. Participants of a scheme in New York's Hudson Valley derive their energy from solar panels installed on their roof tops. When the sun shines brightly, so that the energy produced exceeds usage, they can watch their electricity meter run backwards. To install a small wind turbine, you will need to get a reliable estimate of wind power at the site. The turbine should be mounted as high as possible but still be accessible for maintenance. You may need to get planning permission. Advice on producing your own electricity from the wind is available from the British Wind Energy Association (BWEA) at: http://www.bwea.com/pdf/briefings/smallsystems.pdf

• Buy your next home for its green credentials. Green Moves is one website that advertises homes that are more energy efficient than conventional ones: http://www.greenmoves.com/

• Seek guidance from experts. The Centre for Alternative Technology in Wales (http://www.cat.org.uk/) runs courses on everything from how to make your own biodiesel to installing and testing photovoltaic systems and using wood to heat your house. The EcoHouse, an environmental showhome located in Leicester, in the UK, demonstrates ways in which to make our homes and lives more sustainable. (http://www.environ.org.uk/ecohouse/)

Reducing car usage

• If you own a car, limit your use of it as much as you can. If you commute, offer empty seats to others making the same journey. Where possible, use car-sharing schemes, public transport, a motorbike, bicycle or walk. Two kilograms of carbon can be saved for every journey under three miles for which we walk or use a bicycle.

• Investigate selling your car and using car-sharing schemes instead. An increasing number of companies offer car-pools or lift-sharing options. For example, StreetCar rents out cars from locations across the UK by the hour, day, week or month. To find the nearest car, you simply type in your postcode on the website (www.streetcar.co.uk), and it shows you the 10 nearest locations. Other schemes place neighbouring people with similar journeys together, with the aim of reducing the number of trips made each day with just one person in the vehicle. Londonliftshare is one such example (http://www.londonliftshare.com/).

• If you decide you cannot do without your own car, choose the most fuel-efficient vehicle possible. The fuel consumption of similar-sized cars can vary by as much as 45 per cent. Following the Kyoto Conference on Climate Change in 1997, the European Commission and the European Automobile Manufacturers Association (ACEA) reached a deal in 1998

that committed ACEA members to voluntarily reducing the CO_2 emissions from new passenger cars by a quarter, to an average CO_2 emission figure of 140 grams per kilometre, by 2008. This has led to there being more fuel-efficient vehicles on the market. Fiat has already met the target, Citroën and Renault are on track to meet it and Ford and Peugeot are not far off. Similar agreements exist with Japanese and Korean car-makers. A proposal for new binding rules announced in February 2007 by the European Commission aims to reduce the CO_2 emissions even further, to 130 grams per kilometre by 2012. CO_2 emissions from transport across Europe have been rising rapidly in recent years, growing from 21 per cent to 28 per cent of the total between 1990 and 2004.

Car sharing offers a means to reduce petrol consumption.

In addition to petrol-fuelled cars, it is now possible to buy a variety of alternative fuel cars. These include:

• *Hybrid*: these vehicles have an internal combustion engine as well as an electric motor and battery. They offer lower fuel consumption, as well as reduced CO_2 emissions and potentially fewer local pollutants.
• *Diesel*: these have lower CO_2 emissions per kilometre travelled, but emit more NOx and particles than new petrol vehicles.
• *LPG (Liquified Petroleum Gas) and CNG (Compressed Natural Gas)*: these are generally converted from petrol-fuelled cars. They are usually bi-fuel, so can run on petrol or gaseous fuel. In terms of CO_2 emissions, LPG cars are

TREES AND OFFSETTING CARBON DIOXIDE

Trees use carbon dioxide from the atmosphere, along with sunlight and water, to produce the carbohydrates they need to thrive. The process they use to do this is called photosynthesis. First, water molecules (H_2O) are broken up into oxygen atoms plus hydrogen protons and electrons. The hydrogen is then combined with the carbon and oxygen from carbon dioxide (CO_2) to create new molecules of carbohydrates that are used to build cells. This is what tree biomass is essentially made of. The only waste product is oxygen, which is released back to the atmosphere. Because trees remove carbon dioxide from the atmosphere and lock it away while they are growing, several companies have introduced tree-planting schemes as a means of offsetting carbon released into the atmosphere by driving cars, using energy for lighting and heating, and flying.

Companies such as the Carbon Neutral Company and Climate Care calculate how much carbon a particular activity will generate and then ask for a fee to offset that amount. If you fly from London to Colombo, for example, the Carbon Neutral Company calculates that you will cover 17,468 kilometres and produce 1.9 tonnes of CO_2. It suggests that by paying £14.06 into its natural woodland portfolio, you can off-set the carbon emissions for your flight. The money goes to fund restoration of Northcombe Wood in Devon and Crossroads Wood in Derbyshire, UK. The company also funds forestry projects in Germany, Bhutan, USA, Mexico, Uganda and Mozambique. Climate Care calculates the same flight to Colombo would generate 2.5 tonnes of carbon, which could be offset by contributing £18.78 to a range of its projects, including a rainforest restoration project in Kibale National Park, Uganda.

Since the trees of the first carbon-offsetting project were planted in 1989, the market has become big business. Experts estimate that by the end of the decade it will be worth £300 million. Oil companies, car companies and other carbon-related businesses have been quick to extol the virtues of carbon-offsetting because, they say, it offers a means for us to continue to live fossil-fuel-guzzling lifestyles without feeling guilty about how our actions might affect the global climate. However, as journalist and social critic H.L. Mencken (1880–1956) once said, "For every problem there is a solution which is clean, simple and wrong". The authors of a study linking carbon dioxide and the heat-absorbing effects of trees, published in December 2006, suggest that planting forests to mitigate climate change outside the tropics is not of great use in off-setting emissions, although as trees are a critical component of the Earth's biosphere, action should nonetheless be taken to preserve them.

The scientists, from the global ecology departments at the Carnegie Institution of Washington and the Lawrence Livermore National Laboratory, both in California, USA, found that the forests' usefulness in helping combat global warming depended on their latitude. Because forest canopies are relatively dark, they absorb most of the sun's energy that falls on them and radiate this back out as heat. Pale snowfields, on the other hand, reflect energy from the sun, helping to keep temperatures cooler. The scientists found that between 20° and 50° degrees latitude, forests had a direct warming influence that more or less counter-balanced the cooling effect of removing carbon from the atmosphere. North of 50° latitude, forests warmed the Earth by an average of 0.8°C. Only below 20° latitude did trees cool the planet, by an average of 0.7°C. This is because water evaporating from trees in the humid tropics increases cloudiness, which has a cooling effect.

Even offsetting projects located in the tropics are not an easy panacea for our environmental guilt. Although, in theory, plantations offer a means of temporarily storing carbon, trees are susceptible to disease, fire, timber harvesting, natural decay and the vagaries of local politics.

Forests release CO_2 as well; only a forest that is growing and increasing in biomass will absorb more than it releases. So the sequestration potential of forestry schemes depends on growing and sustaining the forest. The rock band Coldplay received much positive publicity when it announced it would offset carbon emissions from its album *A Rush of Blood to the Head* by planting 10,000 mango trees in Karnataka, India. However, many of the trees died from insufficient water and several villagers who received saplings complained that they did not receive money to cover labour, insecticide or spraying equipment. A combination of prolonged drought and poor organization on the ground were blamed for the failure of the project.

Other tree-planting schemes have impinged on local people's land and livelihoods. The very first carbon offsetting project was conceived by US company Applied Energy Services (AES), together with World Resources Institute, USAID and the NGO CARE, in 1989. At the time AES was seeking approval for a new 183-megawatt coal-fired power plant in Connecticut. It was given the go-ahead thanks to its proposed "mitigation" project, which involved planting 50 million non-native pine and eucalyptus trees on 40,000 smallholdings in the Western Highlands of Guatemala. Sociologist Dr Hannah Wittman of Simon Fraser University in Vancouver later described the project as a "dismal failure", on account of the fact that it criminalized local activities such as gathering firewood, promoted conflicts over the rights to trees, used species that were largely inappropriate for the area and resulted in land degradation. And a decade after the trees were planted, evaluators Winrock International concluded AES's offset target was far below the anticipated level.

Aware of the criticism and shaky scientific knowledge of reforestation schemes for offsetting carbon emissions, some companies have diversified by investing in renewable energy projects in addition to forestry projects. For example, 80 per cent of Climate Care's projects are now made up of sustainable energy projects such as installing treadle pumps to help irrigate land in India, promoting low-wood-use stoves in Madagascar, providing energy-efficient lamps for schools in Kazakhstan, and installing energy-efficient lighting into low-income households in South Africa. Some projects have been praised for the positive impact they have had; others criticized for making very little or no difference. An article in *The Independent* published in 2007 highlighted how treadle pumps, paid for by off-setters through Climate Care, were revolutionizing subsistence farming in India. The simple pumps, which Climate Care is distributing as replacements to polluting diesel versions, draw up groundwater for irrigation from a depth of 10 metres, even in summer. However, one of Climate Care's other projects has been criticized. The scheme provided inhabitants of Guguletu township, South Africa, with 10,000 low-energy light bulbs, but the residents were too poor to be able to replace them once they expired and the company had no process in place to monitor whether their estimated carbon saving was correct. Critics also suggested that they should have aimed the project at wealthier citizens, as people might in hindsight view the low-energy bulbs as the "poor man's light", and decide not to use them.

Finding out whether the scheme you pay into is really making a difference should get easier, as regulation is introduced. In January 2007, the UK announced that it would be the first country to set standards for offsetting companies. But critics say that carbon offsetting is the wrong approach, because what is really required is a change of lifestyle that will cut use of fossil fuels. In other words, we all need to reduce our individual consumption of carbon. And the only guilt-free way to achieve that is by flying less, using public transport instead of private cars, and slashing our heating and electricity use.

usually better than petrol cars, but less efficient than diesel ones. CNG cars have emissions on a par with diesel cars, with other pollutants emitted in similar concentrations to petrol cars.

For more information on cars and fuels available in the UK, see www.vcacarfueldata.org.uk.

Flying less

Aircraft fly at altitudes of between nine and 13 kilometres, emitting gases and particles directly into the upper troposphere and lower stratosphere as they go. The principal emissions are CO_2 and water vapour; other major emissions are nitric oxide (NO) and nitrogen dioxide (NO_2) (which together are termed NOx), sulphur oxides (SOx) and soot. The gases and particles that aeroplanes emit affect the composition of the atmosphere in several ways. They alter the concentration of the greenhouse gases carbon dioxide (CO_2), ozone (O_3), and methane (CH_4); trigger formation of condensation trails; and potentially increase cirrus cloudiness. All these actions contribute to climate change.

Because CO_2 is long-lived in the atmosphere, emissions of this gas from aircraft become well mixed across the world and their effect is indistinguishable from emissions of the same volume from other sources. But the other gases have shorter residence times and remain concentrated near flight routes, primarily in the northern mid-latitudes. Water vapour emitted by planes forms ice crystals in the upper troposphere, generating vapour trails and cirrus clouds, which trap the Earth's heat. Meanwhile, the extra NOx increases ozone in the upper troposphere and lower stratosphere and reduces methane. The extent to which the addition of NOx affects ozone concentrations depends on the altitude at which the plane emits the NOx and varies from regional in scale in the troposphere to global in scale in the stratosphere. Scientists believe that the different ways in which non-CO_2 aviation emissions act on the atmosphere have a "radiative forcing ratio" of around +2.7, which means that the total warming effect of aircraft emissions is 2.7 times greater than their CO_2 emissions alone.

Aviation is currently responsible for around 3.5 per cent of all human-caused climate change emissions. While this seems quite small, it is the fastest-growing source of CO_2 emissions, and this figure is forecast to rise to 15 per cent of the worldwide total by 2050. The number of people flying is likely to double within 15 years. Worldwide, the industry generates nearly as much CO_2 annually as that from all human activities in Africa; a single person flying a return trip between London and New York is responsible for between 1.5 and two tonnes of CO_2. Currently the biggest emitter in terms of aviation emissions is the USA, followed by the UK. A study undertaken by the UK's Tyndall Centre for Climate Change Research demonstrates that the current expansion of the aviation industry around the globe is completely at odds with governments' Kyoto-set carbon-reduction targets.

The UK is a good example. The Government's long-term objective, which will ultimately be made legally binding through the Climate Change Bill (the initial draft of which was published in 2007), is for a 60 per cent cut in British CO_2 emissions from 1990 levels, by 2050. This target corresponds with the aim to stabilize CO_2 concentrations at a level of 550 parts per million by volume (ppmv). However, as Chapter 4 has shown, avoiding dangerous changes to the climate may require stabilizing carbon dioxide concentrations at 450ppmv or lower. Yet the Government's Aviation White Paper, published in December 2003, aimed to cater for a near-trebling of air passengers by 2030. The Tyndall report suggests that, if aviation continues along "business as usual" lines, it would have to account for half the total permissible emissions by 2050 under the 550ppmv target, or 100 per cent of the total, if the aim is to contain concentrations below the 450ppmv target. However, these figures do not take into account the radiative forcing of 2.7. If this is factored into the

We all enjoy going on holiday but travelling by aeroplane greatly increases our carbon footprint.

equation, the percentages leap to 134 per cent for the 550ppmv target and a wholly unachievable 272 per cent for the 450ppmv aim.

In February 2007, the UK Government doubled air passenger duty on flights from the UK, but this is unlikely to put much of a dent in passenger numbers. A study undertaken by researchers at Oxford University suggests that high-income groups, whose emissions are twice the national average, would rather pay more for flights than change their travel habits. The team, from the university's Centre for the Environment, found that the average climate change impact of a person's annual travel was equivalent to 5.25 tonnes of CO_2. And unlike with road vehicles, we cannot rely on technology to offer improvements. Any radical new aircraft designs and new fuels are likely to be decades away from commercialization and large increases in efficiency are unlikely, given the 70 per cent reduction in fuel consumed by jet engines over the past 40 years. So while the thought of giving up newly discovered weekend breaks to the continent and long-haul adventures in the tropics may be unwelcome, aviation has to be targeted, if governments are to have any hope of significantly reducing emissions.

So should we all stop flying? Co-founder of www.responsibletravel.com, Justin Francis, suggests that we should "fly significantly less", but that we should also consider the impact of our actions on those countries that rely on tourism for a large part of their income. In 10 of the world's poorest 50 countries, tourism is significant and growing, he says. And, since many of these countries lack any real alternative to tourism, their primary assets being their cultures and natural environments, stopping the influx of visitors will have a detrimental impact. The question then is, just how much flying is acceptable? In the introduction to his book *Heat, How to Stop the Planet Burning*, which sets out to demonstrate how de-carbonizing can be achieved without radical changes to our lifestyles, George Monbiot writes: "When I come to examine aviation, I discover that there are simply no effective technological solutions: in this chapter I have failed in my attempt to reconcile the luxuries we enjoy with the survival of the biosphere, and I am forced to conclude that the only possible answer is a massive reduction in flights".

If you are planning a holiday, consider taking a trip close to home, or investigate options for going by ferry, train or coach instead of flying. If you must fly, then offset your emissions, but be aware that it is not the answer to a guilt-free holiday (see box on page 76).

Green leisure
• Ditch the gym for a fitness boot camp in your local park.
• Take public transport or share lifts when travelling to events such as concerts or sports matches. Eat home-made food rather than buying takeaway meals in throw-away packaging (see box, right).
• If you must have electrical items, buy energy-efficient stereos/TVs/DVD players and never leave them on standby.
• Take holidays closer to home; fly less or not at all.

Don't be a shopaholic

• Aim to consume less. Everything we buy, from food to electrical goods, has a carbon footprint. The more energy-intensive the production process and the further the item has travelled from its point of manufacture, the greater this carbon footprint is likely to be. One investigation published on the web claims that the carbon footprint of an American cheeseburger is between 4.35 and 7.35 kilograms of CO_2e (carbon dioxide equivalent) per burger, taking into account food production, processing, transportation and methane emissions from beef cows. If each of the 300 million Americans ate

THE ECOLOGICAL IMPACT OF SPORTING EVENTS

Large sporting events, such as football or rugby matches, can have a great ecological impact on the planet. Andrea Collins and her colleagues at the University of Cardiff studied the impact of the 2004 soccer FA Cup Final in the city by assessing the environmental impact of the spectators' travel to the ground, the food they consumed, and the resultant waste. They then converted the energy and resources used, to come up with an "ecological footprint" showing the amount of land required to support the use of those resources. Energy used was represented as the area of forest required to offset the carbon emissions, while food was considered in terms of the amount of farmland required to grow the crops. The match had a massive total footprint of 3,051 hectares, with half of that coming from transport.

The organizers of major sporting events are using such studies to try to reduce their impact. International events have an even greater impact than major national ones; the 2006 Winter Olympic Games in Turin generated 103,500 tonnes of CO_2 in just 16 days. However, the organizers claim that 70 per cent of these emissions were offset by clean energy projects in Eritrea, Mexico and Sri Lanka, and a tree-planting scheme in Kenya (see box: Trees and offsetting carbon dioxide, page 176). In December 2006, a US cycling team sponsored by Kodak claimed to have become the first carbon-neutral professional US sporting team, again by offsetting its emissions.

The organizers of the 2006 FIFA World Cup made efforts for the Finals to be carbon-neutral, through the Green Goal Initiative. The German Football Association invested €500,000 (US$714,250) in an aid programme for Tamil Nadu, a region of India badly affected by the Boxing Day Tsunami of 2004. The money paid for local people to build biogas generation plants, producing fuel suitable for cooking from cow dung. These replaced conventional fuels of kerosene and wood, protecting local forests. The emissions produced by electricity used during the World Cup were "reduced" from an estimated 7,540 tonnes to 2,490 tonnes. The initiative also offset transport emissions from 90,000 tonnes to 73,000 tonnes.

one burger a week, accepting the lower estimate of 4.35 kilograms of CO_2e, that would equate to around 65,250,000 annual tonnes of CO_2e being pumped into the atmosphere, approximately the same as that emitted annually by 6.5 million American Hummer vehicles. And according to the Soil Association, the ingredients of a typical Christmas meal eaten in the UK could have travelled a total of 78,890 kilometres, equivalent to two journeys around the world, releasing 37 kilograms of CO_2.

• Go vegetarian, or better still, vegan. According to the United Nations Food and Agriculture Organization, if land clearance, feed crop production, the energy used by farms and methane emissions are considered, the rearing and slaughter of cows and other animals generates 18 per cent of total human-induced greenhouse gas emissions globally. It is also a major cause of land and water degradation; the sector occupies 30 per cent of

ice-free land on the planet and consumes eight per cent of human water use.

• Buy locally-grown fruit and vegetables rather than choosing produce that has been flown in from warmer climes. Opting to have vegetables delivered as part of a veg-box scheme and buying provisions online can help cut the number of individual car journeys made.

• Re-use bags, rather than accepting plastic ones, and reject items with unnecessary packaging.

Get political

• Support campaigning groups and lobby politicians for change. Meeting the world's greatest challenge will require a monumental effort from governmental leaders and individuals alike.

Sourcing local produce is a simple way to reduce your carbon footprint.

ERA	AGE	SYSTEM	ARRANGEMENT OF LAND AND SEA
Cenozoic	1.8 million years ago to the present day	Quaternary	When the Earth's movements around the sun reduce solar radiation reaching high latitudes, the isolation of the poles from warm equatorial waters means they accumulate heavy coverings of ice. This lowers sea levels, reducing warm-water circulation. The ice sheets reflect more sunlight, further reducing temperatures. Changes in Earth's orbit around the sun on cycles of 100,000, 22,000 and 46,000 repeatedly switch the Earth into and out of ice ages.
	24 million years ago to 1.8 million years ago	Neogene	North and South America finally join together. The western Mediterranean becomes sealed off, prompting the sea to almost dry up. Generally, the world dries out, and large parts of Africa and Asia are taken over by deserts. The continued drift of the continents closes off the waters of the far Arctic and ultimately results in Antarctica becoming isolated and encircled by the continuous Antarctic Circumpolar Current.
	65 million years ago to 24 million years ago	Palaeogene	The components of Gondwana continue to separate, with Africa, Antarctica, South America and Australia all pulling apart from each other and drifting towards their modern-day positions. North America and Asia are intermittently joined by a land bridge; Greenland and North America begin to separate. North America and South America are divided by a sea.

CLIMATIC FEATURE	LIFE
The climate cools and warms with the repeated advance and retreat of polar ice in a series of ice ages and interglacials. The last ice age ends around 10,000 years before the present time, ushering in a relatively stable and warm global climate. This continues until human activities such as deforestation and the burning of fossil fuels push up concentrations of greenhouse gases in the atmosphere and cause global temperatures to rise rapidly from the mid-1900s.	Modern flora and invertebrate species evolve alongside modern humans. Many large mammals, such as the mammoth, sabre-toothed tiger, giant ground sloths and long-horned bison become extinct around 10,000 years before the present time. Early human hunters may be responsible for these extinctions. By the 21st century, over-exploitation of the planet's resources through hunting, fishing and destruction of the world's habitats, coupled with human-induced climatic change, is threatening a vast array of species.
The world's climate cools drastically, possibly because of the uplift of the Himalayas by the Indian subcontinent ramming into Asia.	Modern mammals and flowering plants evolve. The evolution of grass gives rise to long-legged running animals such as horses. In the later Neogene, Hominids appear in the African savannahs.
The climate continues to be warm and tropical around the world. Around 55 million years ago a massive injection of carbon into the atmosphere, from volcanic eruptions or subsea methane stores causes the temperature to rise rapidly by 5-8°C around the world.	Mammal and bird groups diversify and flourish in the tropical temperatures. The continents are isolated by shallow seas, so separate lineages develop on each land mass.

ERA	AGE	SYSTEM	ARRANGEMENT OF LAND AND SEA
Mesozoic	144 to 65 million years ago	Cretaceous	The breakup of Pangea continues. This leads to increased regional differences in floras and faunas between the northern and southern continents.
	206 to 144 million years ago	Jurassic	By early Jurassic times, south-central Asia has become amalgamated. The wide Tethys ocean continues to divide land in the north from Gondwana. From the mid-Jurassic, Pangea begins to break up. In the late Jurassic the Central Atlantic is a narrow ocean dividing Africa from eastern North America. Eastern Gondwana begins to separate from Western Gondwana.
	245-208 million years ago	Triassic	By the Triassic, Pangea is fully amalgamated into a single, giant super-continent on which animals freely wander from pole to pole.

CLIMATIC FEATURE	LIFE
In the early Cretaceous, cool temperate forests cover the polar regions, and snow falls during winter seasons. Then, volcanoes inject high concentrations of CO_2 into the oceans and atmosphere, reducing the amount of oxygen in the seas. Petroleum-rich rocks form in this oxygen-poor environment. The increased CO_2 warms the planet, through the natural greenhouse effect. This is one of the warmest periods in Earth's history; during the late Cretaceous the global climate is warmer than today's and the polar ice melts away.	Life continues in the Cretaceous much as it did in the Jurassic. Dinosaurs inhabit forests of ferns, cycads and conifers. Ammonites, belemnites and fish share the sea with large marine reptiles, while pterosaurs and birds inhabit the skies. The first flowering plants appear and insects diversify. The evolution of the bee is an important event for the development of flowering plants. However, another mass extinction marks the end of the Cretaceous, with 85 per cent of species disappearing including the dinosaurs. A layer of sediment containing iridium now marks this boundary in the geological record. (Some scientists believe a meteorite striking the Yucatan desert in Mexico caused the extinction and the iridium layer. Others suggest massive volcanic outpourings in India and Pakistan were to blame by altering the global climate and changing the chemistry of the oceans.)
Initially, the interior of Pangea remains very arid and hot. Deserts cover what is now the Amazon and Congo rainforests. Meanwhile, China, surrounded by moisture-laden winds is lush and verdant. Later, as Pangea begins to break up, the climate changes. The interior became less dry, and seasonal snow dusts the polar regions.	Dinosaurs dominate the land fauna, the largest being the sauropods, such as Diplodocus and Brachiosaurus. These are herbivores. There are also predatory species such as Allosaurus. The first birds, Archaeopteryx, appear, along with small early mammals. In the seas, icthyosaurs live alongside plesiosaurs, giant marine crocodiles, ammonites, belemnites, sharks and rays. Ferns, ginkgoes, true cycads and conifers dominate the flora.
The interior of Pangea is hot and dry, with warm temperate climes extending to the Poles. This period may represent one of the hottest times in Earth's history. Scientists believe Milankovitch cycles influenced the climate at this time.	Following the great extinction, the surviving species spread and re-colonize. The early Mesozoic is dominated by ferns, cycads, ginkgophytes, bennettitaleans, and other unusual plants. Modern gymnosperms, such as conifers, first appear in their current recognizable forms in the early Triassic. Warm-water faunas spread across the Tethys ocean, which divides the northern land from Gondwana.

ERA	AGE	SYSTEM	ARRANGEMENT OF LAND AND SEA
Palaeozoic	290 to 248 million years ago	Permian	By the beginning of the Permian, the motion of the Earth's crustal plates has fused the major continental land masses together as Pangea. The world essentially comprises vast areas of land and water.
	354 to 290 million years ago	Carboniferous	Vast coal swamps (the source of the hydrocarbons we use for fuel and energy today) characterise the Carboniferous. Ocean covers the entire surface of the globe except for a small, localized section where a super-continent, Pangea, begins to form. Ice sheets repeatedly expand and contract, leading alternately to lower and higher sea levels.
	417 to 354 million years ago	Devonian	Three continental bodies exist. North America and Europe lie together at the equator. Part of modern-day Siberia is located to the north. To the south is a land mass comprising today's South America, Africa, Antarctica, India and Australia.
	443 to 417 million years ago	Silurian	The continents of Laurentia and Baltica collide, closing the northern part of the Iapetus Ocean and forming the 'Old Red Sandstone' continent. Coral reefs expand and land plants begin to colonize the barren continents.
	495 to 443 million years ago	Ordovician	Ancient oceans separate four barren continents named Laurentia (Scotland and North America), Baltica (Scandinavia), Siberia and Gondwana. The Iapetus Ocean divides Laurentia and Baltica from 400 to 600 million years before present.

CLIMATIC FEATURE	LIFE
The interior regions are dry, with great seasonal fluctuations, because there are no nearby water bodies to moderate the climate. As the interiors dessicate, glaciation decreases. Rapid global warming at the end of the Permian creates a 'hot-house' world that leads to the great Permo-Triassic extinction.	The super-heated climate makes life difficult for many species. The largest mass extinction recorded in the history of life on Earth happens at the end of the Permian; 99 per cent of all life on Earth perishes. Most marine vertebrates became extinct at this time. The groups that survive are greatly diminished in number. On land, the extinction of dinapsids and synapsids paves the way for the arrival of the dinosaurs.
At the start of the Carboniferous, the climate is generally tropical and humid throughout the year. Tropical rainforests stretch from Arctic Canada to Newfoundland and Western Europe. Later, the desert regions in mid-North America begin to contract and the Southern Hemisphere begins to cool. Slowly, an ice-cap starts to grow out from the South Pole.	Many spectacular plant forms dominate the land. The evolutionary introduction of the amniote egg makes it easier for vertebrates to colonise the land. The ancestors of birds, mammals and reptiles are now able to reproduce on land without their embryos drying out. Creatures such as bryozoans, brachiopods, sharks and bony fish inhabit the waters.
Several climatic zones exist during the Devonian. There is a narrow equatorial tropical belt, broad subtropical arid zones extending to about 35° latitude, and temperate zones extending to the poles. In the Late Devonian, the southern 'cool temperate' zone expands, and glaciers form in parts of far western Gondwana (today's northern South America).	Land plants diversify; ferns, horsetails and seed-plants appear and produce the first trees and forests. The first land-living vertebrates and arthropods colonize the land. Brachiopods and echinoderms are common in the ocean. New kinds of fish evolve.
The Earth's climate stabilises. Large bodies of ice melt, prompting sea levels to rise.	Coral reefs appear and fishes evolve.
The world's climate remains warm and moist until the mid-Ordovician. But when Gondwana hits the South Pole in the late Ordovician, glaciers form and sea levels drop in response. By the late Ordovician, ice blankets almost all of Gondwana. It is one of the coldest periods in Earth's history.	Marine creatures diversify; a typical community includes graptolites, trilobites, brachiopods, red and green algae, primitive fish and corals. Plants probably invade the land at this time. At the end of the Ordovician, 60 per cent of marine invertebrate genera and 25 per cent of all families become extinct, possibly because of the climate's shift to cool conditions.

ERA	AGE	SYSTEM	ARRANGEMENT OF LAND AND SEA
Palaeozoic	545 to 495 million years	Cambrian	As the Cambrian gets under way, Rodinia begins to fragment into smaller continents (not all corresponding to ones we see today).
Proterozoic	2.5 billion to 543 million years ago	Precambrian	Stable continents appear. Most of the Earth's landmasses collect in a single landmass called Rodinia, 1000 million years before present.
Archean	3.8 to 2.5 billion years ago		The Earth's crust cools until rocks and continental plates begin to form.
Hadean time	4.5 to 3.8 billion years ago		The solar system is forming.

CLIMATIC FEATURE	LIFE
Warmer times follow in the wake of the Precambrian ice. None of the continents are located at the poles, and so land temperatures remain mild. The global climate is probably warmer and more uniform than at the present time.	Major groups of animals appear in fossil record. The number of life forms rapidly diversifies in what is now known as the 'Cambrian explosion'.
Oxygen levels rise from around one per cent of today's levels to more than 15 per cent, on account of expanding cyanobacteria. During the Late Precambrian ice begins to cover the high latitudes.	Cyanobacteria become widespread. They photo-synthesize and produce oxygen as a by product. Bacteria have to evolve to stop the higher levels of oxygen from poisoning them. One method they employ, oxidative respiration, has the advantage of producing large amounts of energy for the cell (it is now widespread in animals and plants). As ice spreads, the decrease in global temperature leads to mass extinctions that eliminate many warm water species.
The atmosphere is composed of methane, ammonia and other gases that would be toxic to us today.	Bacterial microfossils appear – the first sign of life on Earth.

Aars, J., Lunn, N.J. and Derocher, A.E. (Eds.) 2005. Proceedings of the 14th Working Meeting of the IUCN Polar Bear Specialist Group, 20–24 June, Seattle, Washington, USA. *IUCN – The World Conservation Union.*

Adam, D. 11 October 2006. "Water for millions at risk as glaciers melt away", *The Guardian.*

Adam, D. 16 June 2005. "How will carbon capture and storage work?" *The Guardian.*

Adam, D. 22 July 2006. "Drought, gales and refugees: what will happen as UK hots up", *The Guardian.*

Adam, D. 3 February 2007. "Worse than we thought", *The Guardian.*

Adams, J., Maslin, M. and Thomas, E. 10 December 2006. "Sudden Climate Transitions during the Quaternary", (in press with *Progress in Physical Geography*). http://www.esd.ornl.gov/projects/qen/transit.html

Adger, N. et al. 6 April 2007. Climate Change 2007: Climate Change Impacts, Adaptations and Vulnerability. Working Group II Contribution to the Intergovernmental Panel on Climate Change Fourth Assessment Report.

Adler, R. 17 July 1999. "Africa's badlands", *New Scientist*, issue 2195, page 22.

Agrawala, S. (Ed.). 18 January 2007. "Climate change in the European Alps: adapting winter tourism and natural hazards management." Executive summary. *Organisation for Economic Co-operation and Development.*

Alley et al. February 2007. "Climate Change 2007: The Physical Science Basis. Summary for Policymakers." *Contribution of Working Group I to the Fourth Assessment Report of the Intergovernmental Panel on Climate Change.*

Alley, R., Mayewski, P., Peel, D. and Stauffer, B. October 1996. "Twin ice cores from Greenland reveal history of Climate Change, More", *Earth in Space*, vol. 9, No. 2, 12–13.

Alley, R.B. 15 February 2000. "Ice-core evidence of abrupt climate changes", *Proceedings of the National Academy of Sciences of the United States of America*, vol. 97, issue 4, pages 1331–1334.

Amos, J. 21 April 2005. "Antarctic glaciers show retreat", *BBC News Online.*

Arrhenius, S. 1896. "On the influence of carbonic acid in the air upon the temperature of the ground", *The London, Edinburgh and Dublin Philosophical Magazine and Journal of Science*, 41, 237–276.

Asner, G.P., Knapp, D.E., Broadbent, E.N., Oliveira, P.J.C., Keller, M. and Silva, J.N. 21 October 2005. "Selective logging in the Brazilian Amazon", *Science*, vol. 310, No. 5747, pages 480–482.

Associated Press. 23 January 2007. "Smoking gun report to say global warming here", *CNN.com.*

Baillie, S.R., Marchant, J.H., Crick, H.Q.P., Noble, D.G., Balmer, D.E., Coombes, R.H., Downie, I.S., Freeman, S.N., Joys, A.C., Leech, D.I., Raven, M.J., Robinson, R.A. and Thewlis, R.M. 2006. "Breeding Birds in the Wider Countryside: their conservation status 2005." *BTO Research Report* No. 435.

Bakalar, N. 18 January 2006. "Whale Birth Decline Tied to Global Warming, Study Says", *National Geographic News.* NationalGeographic.com.

Beaugrand, G., Reid, P.C., Ibañez, F., Lindley, J.A., and Edwards, M. 31 May 2002. "Reorganization of North Atlantic Marine Copepod Biodiversity and Climate", *Science*, vol. 296, No. 5573, pages 1692–1694.

Benestad R. and Pierrehumbert, R. 17 May 2006. "El Niño and global warming", *Realclimate.org.*

Bilefsky, D. 8 February 2007. "Europe proposes to reduce new cars' carbon dioxide", *New York Times*.

Black, R. 1 November 2005. "Concepts for a low carbon future", *BBC News Online*.

Black, R. 18 November 2004. "Climate change skeptics wrong", *BBC News Online*.

Black, R. 29 December 2006. "Climate 2006. Rhetoric up, action down", *BBC News Online*.

Black, R. 30 January 2006. "Stark warning over climate change", *BBC News Online*.

Black, R. 4 April 2006. "'Major melt' for Alpine glaciers", *BBC News Online*.

Black, Richard. 24 January 2006. "Sea energy 'could help power UK'", *BBC News Online*.

Bows, A., Upham, P. and Anderson, K. "Growth scenarios for EU and UK Aviation: contradictions with climate policy", 16 April 2005. *Report by Tyndall Centre for Climate Change Research and University of Manchester for Friends of the Earth Trust Ltd*.

Brahic, C. 1 February 2007. "Sea level rise outpacing key predictions", *NewScientist.com*.

Brahic, C. 10 November 2006. "Carbon emissions rising faster than ever", *NewScientist.com* .

Brahic, C. 14 December 2006. "2006 was Earth's sixth warmest year on record", *NewScientist.com* .

Brahic, C. 14 December 2006. "Global warming link to hurricanes likely but unproven", NewScientist.com .

Brahic, C. 17 January 2007. "Major sporting events going for green", *NewScientist.com* .

Brahic, C. 22 November 2006. "Emissions of key greenhouse gas stabilise", *NewScientist.com* .

Bréon, F.M. 4 August 2006. "How do aerosols affect cloudiness and climate?" *Science*, vol. 313, No. 5787, pages 623–624.

Briggs, H. 3 December 2000. "Ozone Hole 'set to shrink'", *BBC News Online*.

Brooks, N., Nicholls, R. and Hall, J. 2006. "Sea Level Rise: Coastal Impacts and Responses." *WBGU – German advisory council on global change, Special Report 2006*. http://www.wbgu.de/wbgu_sn2006_ex03.pdf

Bryden, H.L., Longworth, H.R. and Cunningham, S.A. 1 December 2005. "Slowing of the Atlantic meridional overturning circulation at 25°N", *Nature*, vol. 438, pages 655–657.

Byfield, V. 2006. The Atlantic Conveyor. In Burning Ice: Art & Climate change. *Cape Farewell*.

Caldeira, K. 16 December 2006. "Planting trees is far from pointless." Letter in response to "Planting trees to save planet is pointless, say ecologists", *The Guardian*.

Camuffo, D. and Sturaro, G. 2003. "Sixty-cm submersion of Venice discovered thanks to Canaletto's paintings", *Climatic Change* 58, 333–343, *Kluwer Academic Publishers*.

Carey, B. 24 July 2006. "Sahara Desert was lush and populated only temporarily", *FoxNews.com*.

Chamling Rai, Sandeep (Ed.) March 2005. "An overview of glaciers, glacier retreat, and subsequent impacts in Nepal, India and China." *WWF Nepal Program*.

Chandler, D.L. 14 December 2006. "Shorelines may be in greater peril than thought", *NewScientist.com* .

Chandler, D.L. 24 November 2005. "Record ice core reveals Earth's ancient atmosphere", *NewScientist.com* .

Collins, M. January 2005. "El Niño- or La Niña-like climate change?", *Climate Dynamics*, vol. 24, page 89–104.

Coonan, C. 17 November 2006. "Global warming: Tibet's lofty glaciers melt away", *The Independent*.

Copley, J. 16 April 2005. "Sports events leave a giant 'ecological footprint'", *NewScientist.com* .

Cox et al. 2000 Results from carbon cycle experiments. Predictions of accelerated climate change. http://www.metoffice.gov.uk/research/hadley-centre/models/carbon_cycle/results_trans.html

Darwish, A. June 1994. "Water Wars", lecture

at the Geneva Conference on Environment and the Quality of Life.

Dhillon, A. and Harnden, T. 29 April 2006. "How Coldplay's green hopes died in the arid soil of India", *The Telegraph*.

Doyle, A. 16 October 2006. "Antarctic ice collapse linked to man-made greenhouse gases", *ABC News/Reuters*.

Edwards, R. 7 November 2006. "World faces 'dirty, insecure' energy future", *NewScientist.com* .

Enever, A. 10 December 2002. "Bolivian glaciers shrinking fast", *BBC News Online*.

Francis, J. Downloaded 08 February 2007. "Flying in the face of global warming – to fly or not to fly?" *ResponsibleTravel*. http://www.responsibletravel.com/copy/copy101993.htm

Franks, T. 24 October 2005. "Cows make fuel for biogas plant", *BBC News Online*.

Frauenfelder, R., Zemp, M., Haeberli W., and Hoelzle, M. August 2005. "Worldwide glacier mass balance measurements: trends and first results of an extraordinary year in Central Europe", *Ice and Climate News*, No. 6, pages 9-10.

Fry, C. "Plants and our health", *Kew Magazine*, No. 51, page 28.

Fry, C. 1 September 2004. "Need for carbon sink technologies", *BBC News Online*.

Fry, C. 10 November 2006. "Ice study reveals climate change 'seesaw'", *Guardian Unlimited*.

Fry, C. August/September 2005. "Watch the birdie", *Power Engineer*, page 10.

Fry, C. February/March 2006. "Bright Lights, big cities", *Power Engineer*.

García-Herrera, R., Können, G.P., Wheeler, D.A., Prieto, M.R., Jones, P.D., and Koek, F.B. 2 May 2006. "Ship logbooks help analyze pre-instrumental climate", *Eos*, vol. 87, No. 18.

Georgi, F. 21 April 2006. "Climate change hot-spots", *Geophysical Research Letters*, vol. 33.

Glantz, M.H. December 1997. "Climate change – current events", *Our Planet, United Nations Environment Programme (UNEP)*, vol. 9, No. 3,

Goswami, B.N., Venugopal, V., Sengupta, D., Madhusoodanan, M.S,. and Xavier, K. 1 December 2006. "Increasing trend of extreme rain events over India in a warming environment", *Science*, vol. 314, No. 5804, pages 1442–1445.

Graumann, A., Houston, T., Lawrimore, J., Levinson, D., Lott, N., McCown, S., Stephens, S., and Wuertz, D. "Hurricane Katrina: A climatological perspective", Technical report 2005-01. *NOAA's National Climatic Data Center*. http://www.ncdc.noaa.gov/oa/reports/tech-report-200501z.pdf

Green, R.E., Harley, M., Spalding, M. and Zöckler C. (Eds). "Impacts of climate change on wildlife." RSPB, UNEP, WCMC, English Nature and WWF.

Gribben, J. 17 June 1989. "The end of the ice ages? Ice ages have occurred regularly over the past million years – We should soon be starting on the next one, but the greenhouse effect may mean that it never comes." *New Scientist*, issue 1669.

Gribben, J. 1990. Hothouse Earth: The greenhouse effect and Gaia. *Black Swan Books*.

Gribben, J. 2 June 1990. "Methane may amplify climate change", *New Scientist*, issue 1719.

Hamilton, W.L. 7 January 2007. "Incandescence. Yes. Fluorescence, We'll See", *New York Times*.

Handwerk, B. 12 January 2006. "Frog extinctions linked to global warming", *National Geographic News. National geographic.com*.

Hansen, J. 6 December 2005. "Is there still time to avoid 'dangerous anthropogenic interference' with Global Climate?" Transcript of presentation to American Geophysical Union, San Francisco, California. http://www.columbia.edu/~jeh1/keeling_talk_and_slides.pdf

Harden, Blaine. 7 July 2005. "Experts predict polar bear decline: Global warming is melting their ice pack habitat", *Washingtonpost.com.*

Hardin, M and Kahn, R. "Scientific studies of aerosols". Earth Observatory library. http://earthobservatory.nasa.gov/Library/Aerosols/aerosol2.html

Harding, L. 6 December 2006. "Austria's hills aren't alive with the sound of skiing", *The Guardian.*

Harrison, P. (ed). "Global environment outlook yearbook 2006." An overview of our changing environment – *United Nations Environment Programme.*

Hecht, J. 18 June 2005. "Ancient glimpse of seas' bleak future", *New Scientist*, issue 2504, page 19.

Hecht, J. 24 March 2006. "Glacial earthquakes rock Greenland ice sheet", *New Scientist.com* .

Hecht, J. 26 May 2006. "Global warming stretches subtropical boundaries", *New Scientist.com*

Henson, R. September 2006. The Rough Guide to Climate Change. The symptoms, the science, the solutions. *Rough Guides Ltd.*

Herbert, I. and Brown, J. 9 December 2006. "Your carbon footprint revealed: Climate change report finds we produce 11 tons of carbon a year – and breaks down how we do it", *The Independent.*

Hill, M. 21 January 2007. Programs let homes produce green power, *Associated Press.*

His, S. 25 November 2004. "A look at biofuels worldwide." *IFP*

Hoerling, M., Hurrell, J., Eischeid, J., and Phillips, A. 11 November 2005 (2nd revision). "Detection and attribution of 20th century northern and southern African rainfall change", *Journal of Climate.*

Hofmann, D.J. 2006. "Radiative climate forcing by long-lived greenhouse gases: The NOAA Annual Greenhouse Gas Index (AGGI)", *NOAA Earth System Research Laboratory Global Monitoring Division.*

Holmes, B. 10 March 2004. "'Pristine' Amazonian rainforests are changing", *NewScientist.com* .

Hooper, R. 21 January 2006. "Something nasty in the air", *New Scientist*, issue 2535, page 40.

Hooper, R. 9 July 2005. "Sea life in peril as oceans turn acid", *New Scientist*, issue 2507, page 15.

Hopkirk, J. 4 January 2007. "Carbon-offsetting: All credit to them", *The Independent.*

Imbrie, J. and Imbrie, K.P. 1979. Ice ages: solving the mystery. *Harvard University Press.*

Jha, A. 15 December 2006. "Planting trees to save planet is pointless, say ecologists", *The Guardian.*

Johnston, D.C. 8 January 2007. "Taking control of Electric Bill, Hour by Hour", *New York Times Online.*

Jones, P.D., Parker, D.E., Osborn, T.J., and Briffa, K.R. 2006. "Global and hemispheric temperature anomalies – land and marine instrumental records. Trends: A Compendium of Data on Global Change." *Carbon Dioxide Information Analysis Center, Oak Ridge National Laboratory, US Department of Energy, Oak Ridge, Tenn., USA.*

Karas, J. 13 November 1997. "Climate change and the Mediterranean." *Greenpeace report.* http://archive.greenpeace.org/climate/kimpacts/fulldesert.html

Kerr. R.A. 1 July 2005. "Atlantic climate pacemaker for millennia past, decades hence?" *Science*, vol. 309, No. 5731, pages 41–43.

Kirby, A. 19 October 2004. "Water scarcity: a looming crisis?" *BBC News Online.*

Kirby, A. 2 June 2000. "Dawn of a thirsty century", *BBC News Online.*

Kumar, S. 13 January 2007. "Solar cooking system for apartments developed", *The Hindu.*

Kumar, K.K., Rajagopalan, B., and Cane, M.A. 25 June 1999. "On the weakening relationship between the Indian monsoon and ENSO", *Science*, vol. 284, No. 5423, pages 2156–2159.

Kumar, K.K., Rajagopalan, B., Hoerling, M., Bates, G., and Cane, M. 7 September 2006. "Unraveling the Mystery of Indian Monsoon

Failure During El Niño", *Science*, vol. 314, No. 5796, 115–119.

Kuper, R. and Kröpelin, S. 11 August 2006. "Climate-controlled Holocene Occupation in the Sahara: Motor of Africa's evolution", *Science*, vol. 313, No. 5788, pages 803-807.

Kutscher, C.F. (Ed.) 31 January 2007. "Tackling climate change in the US. Potential carbon emissions reductions from energy efficiencies and renewable energy by 2030", *American Solar Energy Society*.

Kutzbach, J.E. and Liu, Z. 17 October 1997. "Response of the African Monsoon to Orbital Forcing and Ocean Feedbacks in the Middle Holocene", *Science*, vol. 278, No. 5337.

Lean, G. 23 July 2006. "One year to save the Amazon", *The Independent*.

Lean, G. 24 December 2006. "Disappearing world: Global warming claims tropical island", *The Independent*.

Lean, G. 7 May 2006. "Ice-capped roof of world turns to desert", *The Independent*.

Lewis, L. 20 November 2006. "Silicon shortage hits solar power hopes", *Financial Times*.

Lohr, S. 29 November 2006. "Energy use can be cut by efficiency, survey says", *New York Times*.

Lovelock, J. 1995. The ages of Gaia: A biography of our living Earth. *Oxford University Press*.

Ma'anit, A. July 2006. "If you go down to the woods today…", *New Internationalist*.

Mackenzie, D. 11 November 2006. "Glimmer of hope for doomed fish", *New Scientist*, issue 2577, page 10.

Madslien, J. 17 January 2006. "Biofuel raises global dilemmas", *BBC News Online*.

Mason, B. 22 December 2001. "End of an empire? Blame it on the weather", *New Scientist*, issue 2322, page 11.

Matheson, I. 22 April 2005. "Kenya looks underground for power", *BBC News Online*.

McAvan, L. 27 October 2006. Letters: "Spreading impact of climate change", *The Guardian*.

McGuire, B. 27 May 2006. "Climate change: Tearing the Earth apart?" *New Scientist*,

issue 2553, page 32.

McKee, M. 23 September 2004. "Antarctic glaciers slipping faster into the sea", *NewScientist.com*

Mckee, M. 27 October 2004. "Sunspots more active than for 8,000 years", *NewScientist.com* .

Mckenna, P. 14 September 2006. "Winter Arctic sea ice in drastic decline", *NewScientist.com*

Mclaughlin, K. 21 August 2006. "Red tide's worst bloom this year: bad PR", *HeraldTribune.com*.

Menzel A., et al. October 2006. "European phenological response to climate change matches the warming pattern", *Global Change Biology*, vol. 12, issue 10, page 1969.

Merali, Z. 12 January 2006. "The lungs of the planet are belching methane", *New Scientist*, issue 2534, page 13.

Meyer, A. 18 May 2006. "Viewpoint: The fair choice for climate change", *BBC News Online*.

Miller, D.L., Mora, C.I., Grissino-Mayer, H.D., Mock, C.J., Uhle, M.E. and Sharp, Z. 19 September 2006. "Tree-ring isotope records of tropical cyclone activity", *Proceedings of the National Academy of Sciences of the United States of America*, vol. 103, No. 49, pages 14294–14297.

Milmo, C. 29 January 2007. "World faces hottest year ever as El Niño combines with global warming", *The Independent*.

Monbiot, G. 19 December 2006. "Preparing for take-off."
http://www.monbiot.com/archives/2006/12/19/preparing-for-take-off/#more-1036

Monbiot, G. 2006. Heat. How to stop the planet burning. *Allen Lane (Penguin Books)*.

Multiple authors. 2000. Intergovernmental Panel on Climate Change Special Report: Emissions Scenarios. A Special Report of IPCC Working Group III.

Multiple authors. September 2005. *Carbon Dioxide Capture and Storage: Summary for Policymakers. A special report of working group III of the Intergovernmental*

Panel on Climate Change.

Mulvey, S. 7 February 2007. "No end yet to EU car CO_2 fight", *BBC News Online.*

Ng, W-S and Mendelsohn, R. 2005. "The impact of sea level rise on Singapore". Environment and Development Economics, issue 10: pages 201–215. *Cambridge University Press.*

Overpeck, J.T., Otto-Bliesner, Bette L., Miller, G.H., Muhs, D.R., Alley, R.B., and Kiehl, J.T. 24 March 2006. "Palaeoclimatic evidence for future ice-sheet instability and rapid sea-level rise", *Science*, vol. 311, No. 5768, pages 1747–1750.

Pacala, S and Socolow, R. 13 August 2004. "Stabilization wedges: solving the climate problem for the next 50 years with current technologies", *Science*, vol. 305, No. 5686, pages 968–972.

Pain, S. 6 December 2003. "An officer and a weatherman", *New Scientist*, issue 2424, page 40.

Parkinson, D. 10 January 2007. "States, provinces will follow California's lead on carbon emissions: CIBC", *Globe and Mail.*

Pearce, F. 11 August 2005. "Climate warning as Siberia melts", *NewScientist.com.*

Pearce, F. 15 September 2005. "Warming world blamed for more strong hurricanes", *NewScientist.com.*

Pearce, F. 18 September 2002. "Africa's deserts are in 'spectacular' retreat", *NewScientist.com.*

Pearce, F. 2 May 1998. "Wind of change", *New Scientist*, issue 2132, page 35.

Pearce, F. 2 September 2000. "Feel the pulse", *New Scientist*, issue 2254, page 30.

Pearce, F. 21 September 2002. "Africans go back to the land as plants reclaim the desert", *New Scientist*, issue 2361, page 4.

Pearce, F. 22 June 2006. "Kyoto promises nothing but hot air", *New Scientist*, issue 2557, page 10.

Pearce, F. 27 August 2005. "Global warming: the flaw in the thaw", *New Scientist*, issue 2514, page 26.

Pearce, F. 30 April 2005. "Squeaky clean fossil fuels", *New Scientist*, issue 2497, page 26.

Penman, D. 22 September 2003. "First power station to harness moon opens", *NewScientist.com.*

Penner, J.E., Lister, D.H., Griggs, D.J., Dokken, D.J., and McFarland, M. (Eds.) 1999. "Aviation and the Global Atmosphere: Summary for Policymakers", a Special Report of IPCC Working Groups I and III in collaboration with the Scientific Assessment Panel to the Montreal Protocol on Substances that Deplete the Ozone Layer.

Peterson, B.J., Holmes, R.M., McClelland, J.W., Vörösmarty, C.J., Lammers, R.B., Shiklomanov, A.I., Shiklomanov, I.A., and Rahmstorf, S. 13 December 2002. "Increasing River Discharge to the Arctic Ocean", *Science*, vol. 298, No. 5601, 2171–2173.

Peterson, L.C. and Haug, G.H. July/August 2005. "Climate and the collapse of Maya civilization: A series of multi-year droughts helped to doom an ancient culture", *American Scientist Online.* http://www.americanscientist.org/template/AssetDetail/assetid/44510?fulltext=true&print=yes

Pickrell, J. 25 October 2006. "Global warming fuels fungal toad-killer", *NewScientist.com.*

Pickrell, J. 4 September 2006. "Instant Expert: Hurricanes", *NewScientist.com* .

Pittock, B. (Ed.) 2003. "Climate change – An Australian Guide to the Science and Potential Impacts." Australian Government Department of the Environment and Heritage Australian Greenhouse Office. http://www.greenhouse.gov.au/science/guide/

Pope, V. Downloaded 8 February 2007. "Models 'key to climate forecasts'", *BBC News Online.*

Prasad, R. 9 December 2006. "In the rice paddies of Sri Lanka, a new enemy: salt", *The Guardian.*

Press Association, 1 August 2006. "Last month was hottest since UK records began", *The Guardian.*

Rahmsdorf, S. 2006. "Thermohaline Ocean Circulation", *Encyclopedia of Quaternary Sciences*. Elsevier.

Rahmstorf, S. "Abrupt Climate Change", *Weather catastrophes and climate change: the state of science*. Munich Re.

Rahmstorf, S. 2001. "Climate: Abrupt change." http://www.pik-potsdam.de/~stefan/Publications/Book_chapters/abrupt.pdf

Randerson, J. 27 October 2006. "Sea Change: why global warming could leave Britain feeling the cold", *The Guardian*.

Ravilious, K. 18 January 2007. "2100: A world of wild weather", *NewScientist.com*.

Readinger, C. February 2006. "Ice core proxy methods for tracking climate change", *CSA Discovery Guides*.

Reddy, T. July 2006. "Blinded by the light", *New Internationalist*.

Reid, J. 28 February 2006. "Water wars: climate change may spark conflict", *The Independent*.

Rekacewicz, P. (cartographer) 2000. "Potential impact of sea-level rise on Bangladesh", UNEP/GRID-Arendal Maps and Graphics Library. http://maps.grida.no/go/graphic/potential_impact_of_sea_level_rise_on_bangladesh

Rincon, P. 25 July 2006. "Wind power dilemma for Lewis", *BBC News Online*.

Roach, J. 10 August 2006. "Greenland ice sheet is melting faster, study says", *National Geographic News*.

Rodhe, H and Charlson, R. (Eds.) 1998. "In commemoration of Svante Arrhenius: scholar, and teacher with a global perspective." In *The legacy of Svante Arrhenius understanding the greenhouse effect*. Royal Swedish Academy of Sciences and Stockholm University.

Romero, S. 2 February 2007. "Bolivia's only ski resort is facing a snowless future", *New York Times*.

Ruddiman, W. F. 2005. Plows, Plagues and Petroleum: How Humans took Control of Climate. Princeton University Press.

Russell, B. and Morris, N. 28 February 2006.

"Armed forces are put on standby to tackle threat of wars over water", *The Independent*.

Sample, I. 14 October 2004. "Pressure points", *The Guardian*.

Sample, I. 30 June 2005. "The father of climate change", *The Guardian*.

Samuelson, R.J. 8 February 2007. "The dirty secret about global warming", *MSNBC*. http://www.msnbc.msn.com/id/17025081/site/newsweek/print/1/displaymode/1098/

Schellnhuber, J. 26 August 2004. "Switches and choke points in the Earth system: the planet's Achilles' heels." Presentation at EuroScience Open Forum in Stockholm.

Schmidt, G. September 2004. "Methane: A Scientific Journey from Obscurity to Climate Super-Stardom", NASA Goddard Institute for Space Studies and Center for Climate Systems Research, Columbia University in New York. www.giss.nasa.gov/research/features/methane/

Schneider, S. H. and Lane, J. February 2005. "An Overview of 'Dangerous' climate change", presented at Avoiding Dangerous Climate Change Conference, Exeter.

Seagar, A. 27 November 2006. "How mirrors can light up the world", *The Guardian*.

Selley, R.C. The winelands of Britain: past, present and prospective. Petravin.

Shepard, A. 13 January 2007. "Eco-worrier: You snow it makes sense", *Times Online*.

Shukman, D. 28 July 2004. "Greenland ice-melt 'speeding up'", *BBC News Online*.

Sincell, M. 9 July 1999. "New Clue to Sahara's Origins", *Science*. http://sciencenow.sciencemag.org/cgi/content/full/1999/709/1

Slavin, T. 3 January 2007. "Leading by example", *The Guardian*.

Srivastava, S. 4 January 2007. "India's airlines look to fly high", *Asia Times online*.

Staff and agencies. 10 January 2007. "Call for 20% EU cut in greenhouse gas emissions", *The Guardian*.

Staff and agencies. 16 January 2007. "UK to set world's first carbon offsetting standards",

Guardian Unlimited. http://environment.guardian.co.uk/print/0,,329687103-121526,00.html

Staff writers. 3 May 2006. "Tibetan glacier melt leading to sandstorms in China", *News about planet Earth*. Terradaily.

Steinfeld, H., Gerber, P., Wassenaar, T., Castel, V., Rosales, M., and de Haan, C. 2006. "Livestock's long shadow. Environmental issues and options", The Livestock, Environment and Development Initiative and the Food and Agriculture Organization of the United Nations.

Stern, N. October 2006. *Stern Review: The Economics of Climate Change*.

Stoll, H.M. 1 June 2006. "The Arctic tells its story", *Nature*, vol. 441, News and views, pages 579–581.

Tady, M. 7 December 2006. "Meat contributes to climate change, UN study confirms", *The New Standard*. http://newstandardnews.net/content/index.cfm/items/3956

Tenywa, G and Alweny, S. 14 November 2006. "Return Our Forest", *Panos Features*.

Topping, J.C. and Frey, E. September 2006. "A moral and profitable path to climate stabilization." Overview of the Washington Summit on Climate Stabilization. *Climate Institute*.

Tremlett, G. 11 September 2006. "Global warming to wash away beaches, warns Spanish study", *The Guardian*.

Trenberth, K.E. and Shea D.J. 27 June 2006. "Atlantic hurricanes and natural variability in 2005", *Geophysical research letters*, vol. 33.

Tripathi, R.D. 5 September 2004. "Solar plan for Indian computers", *BBC News Online*.

Trusca, Vlad. 7 December 2004. "Sawdust 2000", Joint Implementation Project. Presentation COP-10, Buenos Aires.

Turner, J., Lachlan-Cope, T.A., Colwell, S., Marshall, G.J., and Connolley, W.M. 31 March 2006. "Significant warming of the Antarctic winter troposphere", *Science*, vol. 311, No. 5769, pages 1914–1917.

Tyndall, J. 1863. "On Radiation Through the Earth's Atmosphere", *The London, Edinburgh, and Dublin Philosophical Magazine and Journal of Science*, 4, 200–207.

Velicogna, I. and Wahr, J. 24 March 2006. "Measurements of Time-Variable Gravity Show mass loss in Antarctica", *Science*, vol. 311, No. 5768, pages 1754–1756.

Visbeck, M. "North Atlantic Oscillation", http://www.ldeo.columbia.edu/NAO/

Wara, M. 8 February 2007. "Is the global carbon market working?" *Nature* 445, 595–596.

Watson, R.T. et al. "Climate Change 2001: Working Group I. The Scientific Basis." IPCC. www.grida.no/climate/ipcc_tar/wg1/041.htm

Watts, J. 8 November 2005. "China pledges to double reliance on renewable energy by 2020", *The Guardian*.

Weart, S. June 2006. "General circulation models of the atmosphere." in *The discovery of global warming*. http://www.aip/history/climate/GCM.htm

White, K.S. et al. 2001. "Climate change 2001: Impacts adaptation, and vulnerability. Technical Summary", Report of Working Group II of the Intergovernmental Panel on Climate Change.

Wigley, T.M.L. 20 October 2006. "A Combined Mitigation/Geoengineering approach to Climate Stabilization", *Science*, vol. 314, No. 5798, pages 452–454.

Wintour, P. 11 December 2006. "Miliband plans carbon trading 'credit cards' for everyone", *Guardian Unlimited*.

Young, E. 10 June 2006. "Raiders of the lost storms", *New Scientist*, issue 2555, page 44.

Young, E. 29 July 2006. "No Sahara Desert, no Egyptian Dynasty", *New Scientist*, issue 2562, page 16.

Young, J. 16 December 2005. "Hurricane Katrina: Plans, decisions and lessons learned", *Voice of America*. http://www.voanews.com/english/archive/2005-09/2005-09-16-voa63.cfm

Zeng, N., Yoon, J.-H., Marengo, J. A., Subramaniam, A., Nobre, C.A., and Birkett, C.M. 13 August 2006. "Causes and Impacts of the 2005 Amazon Drought", *Science*.

Sundry documents

15 October 1998 "Can a Chinese herb win the malaria war?" *BBC News Online.*

2000. "Aviation and global climate change." Report published by Friends of the Earth, Aviation Environment Federation, National Society for Clean Air and Environmental Protection, and HACAN Clearskies. http://www.foe.co.uk/resource/reports/aviation_climate_change.pdf

13 November 2001 "Carbon trading market expands to Chicago, Mexico City", *Environment News Service.*

18 July 2002 "Alaskan glaciers melting faster", *BBC News website.*

30 October 2002. "India rejects climate change pressure", *BBC News Online.*

April 2003 "Climate change and tourism", proceedings of the first International Conference on Climate Change and Tourism. World Tourism Organization.

9 September 2003 "Floods damage ancient Timbuktu", *BBC News Online.*

2003 "LME 29: Benguela Current", National Oceanic and Atmospheric Administration. US Department of Commerce. http://na.nefsc.noaa.gov/lme/text/lme29.htm

February 2004 "Country Analysis Briefs – India: Environmental Issues", *USA Energy Information Administration.*

27 March 2004 "A mirror to cool the world", *New Scientist*, issue 2440, page 26.

March 2004 "Impacts of summer 2003 heat wave in Europe", *Environment Alert* Bulletin 2. United Nations Environment Programme. http://www.grid.unep.ch/product/publication/download/ew_heat_wave.en.pdf

2 August 2004 "Locust emergency in scared Gambia", *BBC News Online.*

9 September 2004 "Hurricane Ivan blasts Caribbean", *BBC News Online.*

10 October 2004 "The Durban Declaration on Carbon Trading."

17 November 2004 "2004 IUCN Red List of Threatened Species Reveals 15,589 Species at Risk of Extinction." News release from Conservation International, http://www.conservation.org/xp/news/press_releases/2004/111704.xml

22 December 2004 "How do we know that recent CO_2 increases are due to human activities? Real Climate: climate science from climate scientists." http://www.realclimate.org/index.php?p=87

25 February 2006 "Greenland's water loss has doubled in a decade", *New Scientist*, issue 2540, page 20.

14 March 2005 "Himalayan glaciers 'melting fast'", *BBC News Online.*

16 May 2005 "Half 'opposed to nuclear power'", *BBC News Online.*

June 2005 "Joint science academies' statement: Global response to climate change." *Key World Energy Statistics 2006*, International Energy Agency.

May 2005 "Avoiding dangerous climate change", International symposium on the stabilisation of greenhouse gas concentrations.

1–3 February 2005 Report of the International Scientific Steering Committee.

4 July 2005 "Climate change: the big emitters", *BBC News Online.*

6 August 2005 "Warming shock written in ice", *New Scientist*, issue 2511, page 17

12 August 2005 "Warning issued over toxic algae", *BBC News Online.*

22 August 2005 Guide to Mediterranean heatwave. *BBC News Online.*

21 September 2005 Energy Action Plan II.

State of California.
http://www.energy.ca.gov/energy_action_plan/2005-09-21_EAP2_FINAL.PDF

26 September 2005 "Infectious disease and dermatologic conditions in evacuees and rescue workers after Hurricane Katrina – multiple states August-September 2005", Centers for Disease Control and Prevention.
http://www.cdc.gov/mmwr/preview/mmwrhtml/mm54d926a1.htm

28 November 2005 "Clean coal technology: how it works", *BBC News Online.*

December 2005 "Climate change and the Greenhouse effect, a briefing from the Hadley Centre", *The Met Office Hadley Centre.*

10 December 2005 "Last-minute climate deals reached", *BBC News Online.*

1 January 2006 "Climate report: the main points", *BBC News Online.*

11 January 2006 "More Brazil cars run on alcohol", *BBC News Online.*

4 February 2006 "Top climatologist forecasts swift ice cap collapse", *New Scientist*, issue 2537, page 7.

15 February 2006 "Flooding in Southern Algeria", *Visible Earth catalogue*, NASA.
http://visibleearth.nasa.gov/view_rec.php?id=20492

18 February 2006 "Sands of time blown back to reveal ancient Sahara", *New Scientist*, issue 2539, page 22.

23 February 2006 "Two new Munich Re publications". News release from Munich Re,
http://www.munichre.com/

28 February 2006 "Carbon 2006: Towards a truly global market. Released at Point Carbon's 3rd Annual Conference: Carbon Market Insights 2006 in Copenhagen"

2 March 2006 "NASA mission detects significant Antarctic ice mass loss." News release from NASA Jet Propulsion Laboratory California Institute of Techology,
http://www.jpl.nasa.gov/news/news.cfm?release=2006-028

30 March 2006 "Rapid temperature increases above the Antarctic". News Release from

British Antarctic Survey, http://www.antarctica.ac.uk/News_and_Information/Press_Releases/story.php?id=281

1 May 2006 "NOAA issues Greenhouse Gas Index", *NOAA News Online.*

6 May 2006 "No winner in future climate league", *NewScientist.com.*

May 2006 "An operator's guide to the EU Emissions Trading Scheme: The steps to compliance", Defra.

21 June 2006 "Making Sweden an oil-free society." Report by the Commission on Oil Independence for the Swedish Government

23 June 2006 "Backing for 'hockey stick' graph", *BBC News Online.*

28 June 2006 "Fertilizers give the lungs of the planet bad breath", *NewScientist.com.*

7 July 2006 "Hydropower rebounds slightly", Worldwatch Institute overview.
http://www.worldwatch.org/node/4245

12 July 2006 "Energy review urges barrage study", *BBC News Online.*

4 August 2006 "Amazonian drought", *Realclimate.com.*

4 August 2006 "Cape Town signs wind power deal", *BBC News Online.*

14 August 2006 "US wind energy installations reach new milestone", American Wind Energy Association.

18 August 2006 "Rising sea levels threaten to wash away entire nations", WWF.
http://www.panda.org/about_wwf/what_we_do/climate_change/problems/impacts/sea_levels/index.cfm

23 August 2006 "Ozone hole stable, say scientists", *BBC News Online.*

29 August 2006 "Drought affects 11.13m hectares of farmland", Chinese Government's official web portal, *Gov.cn.*

29 August 2006 News release: "Treat yourself to a guilt-free, organic Christmas feast with all the trimmings", The Soil Association,
http://www.soilassociation.org/web/sa/saweb.nsf/848d689047cb466780256a6b00298980/a06d0fdf93ee0cbf802571d50062aba4!OpenDocument

7 September 2006 "Greenhouse gas bubbling from melting permafrost feeds climate warming", *Science Daily,* http://www.sciencedaily.com/releaes/2006/09/060907102808.htm

9 September 2006 Precision climate modeling forecast by ORNL researchers. *Science Daily,* http://www.sciencedaily.com/releases/2006/09/060908194011.htm

19 September 2006 "Tree rings could settle global warming hurricane debate", *Mongabay.com,* http://news.mongabay.com/2006/0919-hurricanes.html

20 September 2006 "Power from pig waste project wins CDM approval", news release from Defra.

20 September 2006 "Agreement to acquire through a merger agreement the outstanding issued share capital in Chicago Climate Exchange, Inc", news release from Chicago Climate Exchange.

30 September 2006 "Alarming surge of a potent greenhouse gas", *New Scientist,* issue 2571, page 6

September 2006 "Renewables in Global Energy Supply: An IEA Fact Sheet." International Energy Agency.

4 October 2006 "Arctic sea ice melt accelerating", Environment , http://www.ens-newswire.com/ens/oct2006/2006-10-04-02.asp

16 October 2006 "First direct evidence that human activity is linked to Antarctic ice collapse", British Antarctic Survey news release.

26 October 2006 "Ozone hole reaches record proportions", *New Scientist,* issue 2575, page 6.

27 October 2006 "Climate change – Impacts at 2°C and 3°C", WWF Factsheet.

28 October 2006 "Drought could wipe out 35% of income from grain", *The Guardian.*

31 October 2006 "Climate change fight can't wait", *BBC News Online.*

October 2006 "The Global Carbon Cycle", Unesco-Scope Policy Briefs No. 2.

2 November 2006 "Bulbs must be efficient 'by 2009'", *BBC News Online.*

9 November 2006 "Reef warns of sea level rise", *Science Daily.* http://www.sciencedaily.com/releases/2006/11/061109094732.htm

15 November 2006 "Climate change blamed for India's monsoon misery", *NewKerala.com.* http://www.newkerala.com/news4.php?action=fullnews&id=51218

November 2006 "Marine climate change impacts annual report card 2006." Marine Climate Change Impacts Partnership.

November 2006 "The state of Greenhouse gases in the atmosphere using global observations through 2005." WMO Greenhouse Gas Bulletin 2:1.

13 December 2006 "'Zero carbon' homes plan unveiled", *BBC News Online.*

19 December 2006 "Green light for world's largest windfarm", *The Scotsman.*

20 December 2006 "EU tackles aircraft CO_2 emissions", *BBC News Online.*

20 December 2006 "Q&A: Europe's carbon trading scheme", *BBC News Online.*

21 December 2006 "Climate Change vs Mother Nature: Scientists reveal that bears have stopped hibernating", *The Independent.*

2006 "Energy efficiencies: pipe-dream or reality?" *World Energy Council Statement 2006.*

2006 "Water scarcity, risk and vulnerability." *Human Development Report 2006.*

2006 "GHG Data 2006: Highlights from Greenhouse Gas (GHG) Emissions Data for 1990–2004 for Annex I parties." Submitted under the United Nations Framework Convention on Climate Change.

8 January 2007 Designated National Authority for the CDM (UK DNA): List of projects with UK approval of participation. Defra.

10 January 2007 "EU plans industrial revolution", *BBC News Online.*

1 February 2007 "Taxes 'fail to curb travel CO_2'", *BBC News Online.*

Downloaded 9 January 2006 Drought simulation. Rainfall Exclusion Experiment.

Woods Hole Research Centre.
http://www.whrc.org/southamerica/drought_si
m/index.htm
Downloaded 27 January 2007 "Land-use,
land-use change and Forestry (LULUCF)",
http://unfccc.int/methods_and_science/lulucf/i
tems/1084txt.php
Downloaded 27 January 2007 "A summary
of the Kyoto Protocol."
http://unfccc.int/kyoto_protocol/background/i
tems/2879.php
Downloaded 29 January 2007 "Kyoto
Protocol."
http://unfccc.int/kyoto_protocol/background/i
tems/3145.php
Downloaded 8 February 2007 "Aviation and
Global Climate Change", Friends of the Earth.
http://www.foe.co.uk/resource/reports/avia-
tion_climate_change.pdf
Downloaded 12 February 2007 "Biodiversity
and climate change", United Nations
Environment Programme World Conservation
Monitoring Centre. http://www.unep-
wcmc.org/climate/impacts.htm
Downloaded 14 February 2007 "2005
Hurricane Archive: Hurricane Katrina August

23–30 2005", National Aeronautics and Space
Administration. http://daac.gsfc.nasa.gov/hur-
ricane/2005HurricaneArchive.shtml
Downloaded 15 February 2007 "Fish: the
'inner oceans–outer space' connection",
South Pacific sea-level and climate monitoring
project. http://www.bom.gov.au/pacificsealev-
el/pdf/AMSAT_fisheries_factsht.pdf
Downloaded 22 February 2007
"Catastrophic impact of world's largest
onshore wind farm revealed in maps", Royal
Society for the Protection of Birds (RSPB).
http://www.rspb.org.uk/scotland/action/lewis/
news/maps.asp
Downloaded 22 February 2007 "Climate
Change Information Sheet 18." The Climate
Change Convention, UNFCCC.
http://unfccc.int/essential_background/back-
ground_publications_htmlpdf/climate_change
_information_kit/items/302.php

Index

Page numbers in *italics* refer to captions/illustrations

Aborigines 91
Adhémar, Joseph Alphonse 14, 19
aerosols 45-6
Africa:
  agriculture 116-17
  sea levels 115
  southern: flooding 8
Agassiz, Lake 27
Agassiz, Louis 13-14, 19-20
Ages of Gaia: A Biography of our Living Earth 93
agriculture 23-4, 35, 113-20, 125, 126, 127, 132
  biomass crops 157
aircraft:
  emitting water vapour 44
  reducing use of 178-80
Alaska:
  flooding 8
  glaciers 78
algae: marine 92
Alps 80
Amazon rainforest 28, 30, 68, 69-70
Amazon River 67-8, *68*
ammonium 24
amphibians 84, 87-90
Antarctic Bottom Water 26
Antarctic Circumpolar Current 28
Antarctic Peninsula 74-5
Antarctica 73
  glacier retreat 74-5, 78
  ice 22-3, 24
  ice sheets 73, 75, 83, 95
  Larsen B Ice Shelf 28, 32, 75, *75*
  melting of ice sheet 28, 31, 32
  ozone hole 29, 43
  penguins 83-6
anticyclones 54-5
Arakawa, Akio 108
Arctic 73-4
  sea temperatures 98
  wildlife 84
Arctic Coring Expedition (ACE) 98-9
Arctic Ocean: sea ice 73-4
Arrhenius, Svante 9, 20-1, *20*, 22
Asia:
  agriculture 117

sea levels 115
atmosphere 36, 45
  absorbing heat 20-1
atoms 17
Australasian Palaeohazards Research Unit 102
Australia:
  agriculture 117-19
  climate changes 72
  drought 8
aviation industry 132-3, *132*

Bangladesh 32, 33, 111, *112*, 115
bears 9, 84, 85, 90-1
Beddington Zero-Energy Development (BedZED) 168-9
Bellotto, Bernardo 103
Benz, Karl 20
Bergamaschi, Peter 133
biodiversity 122
biofuels 168
biomass 156-9
birds:
  effects of climate change on 83-7, 88, 89, 121-2
  impact of wind turbines on *154*, 155
blindness 29
Bolivia: glaciers 78
Bond, Gerard 25
Brazil 69
  biofuels 157-8
  carbon emission reductions 152
Britain: wild vines 104-5
British Antarctic Survey 32, 74, 76, 83
British Trust for Ornithology 89
Bryan, Kirk 108
butterflies 84, 85, 121

cacti *71*
Calanus 81
calcium 24
calcium bicarbonate 92
calcium carbonate 22, 63-4, 92, 101
California: seabirds 86-7
California Cooperative of Oceanic Fisheries Investigations 86
Callendar, Guy 21
Canaletto 103
cancer: skin 29
carbon:

capturing 164
conversion into CO_2 49
cycle 39-40, *40*
emissions:
  by industry *131*
  national reductions 152-3
  reducing 172-83
isotopes 47-9
stored in rainforests 66, 67
carbon allowance 169
carbon credits 134, 135
carbon dioxide (CO_2) 39-40, 96-7
  absorbing heat 20
  absorption 40
    by oceans 63
  amount generated 167
  capturing 164
  concentration in atmosphere 8, 9-10, 21, 40, *41*, 129
  dangerous levels 141
  effect on temperatures 20, 22, 141
  emissions 129
    cuts required 144
    global budget 143-4
    setting ceiling 143-4
    targets 129-36, 140-5
  from human activities 40, 48-9
  GWP 48
  increasing long-wave infrared radiation 21
  levels 23, 128-9
  in oceans 22
  offsetting with trees 176-7
  rainforests and 66-7
  released into atmosphere 48
  remaining in atmosphere 10, 40
  role in ice ages 20-1
  weight 48
  world's biggest emitters 145-8
carbon footprint 170-1
  calculating 167
  elements contributing to 171
  reducing 171-83
carbon "sinks" 66, 130, 136
carbonate conpensation depth (CCD) 22
carbonic acid 63

Caribbean Islands: carbon emission reductions 152
cars:
alternative fuel 175-8
reducing use of 174-8
Central America: agriculture 119
Centre for Ecology and Hydrology 88
CFCs 29, 43, 44
Chicago: Community Energy Cooperative 166
China:
  drought 8, 13
  glaciers 79
  greenhouse gas emissions 144-5
chlorine 24, 97
Chrétien, Jean *130*
cities:
on coastal plains 32
temperatures near 46-7
clathrates 29
Clean Development Mechanism 134-5
Cleveland, Cory 67
climate:
  changes 12-50
  dangerous: avoiding 140-4
  human cost 34-5
  human influences 8, 31, 32, 46-50, 52-93
  mitigating impacts of 128-64
  natural causes 13, 47-8
  politics timeline 137-9
  predicting outcomes 94-127
  regional responses to 110-11
  tipping points 28-9, 30
cycles 60-3
effect of oceans on 25-6, 30, 57-61
effects on wildlife 80-91
processes driving 52-93
records 102
regulated by algae 92
Climatological Database for the World's Oceans 102
cloud forest 84, 87
clouds 38
  formation 45
  noctilucent 36
  seeding 164
CO_2 *see* carbon dioxide
coal-burning 21

Combes, Jean 88
combined heat and power (CHP) units 166
computers: predicting future with 105-6, 107-9
Conrad, Timothy 14
consumers: carbon footprint 181-3
Continuous Plankton Recorder (CPR) 81
cooling agents 45-6
corals 49, 101-2, *101*, 113
Coriolis force 53-4, 56, 57
Costa Rica 84, 87
Croll, James 14
cyclones 54-5

Daisyworld 92
Dansgaard/Oeschger (D/O) events 24, 25, 26
Day after Tomorrow (film) 27
Denmark: carbon emission reductions 143
desertification 67
deserts 70-3
diatoms 98
diseases 10, 121
doldrums 54
Domesday Book 159
drought 8, 13, 31, 33, 34, 35, 62, 67-9, 99, 129
drugs 122
Dryas octopetala 26-7
drylands 70-3
dust 97

Earth:
  cooling 164
  magnetic field 18
  orbit round Sun 13, 14-17, *15*
    changes in 15-17, 18-19
    shape of 16
  processes driving 52-93
  timelines 184-91
earthquakes 76-7
ecological footprint 169
economy 33, 124
ecosystems 122, 125, 126, 127
Einstein, Albert 151
Ekström, Göran 76
El Niño 28, 56, 60-3, *62*, 67-8, 69, 86, 119
El Niño/Southern Oscillation (ENSO) 60, 62, 67
electromagnet radiation *38*
Emissions Trading 134, 135-6
energy:
  conservation 166-70
  demand for 165-6

production 66, 149-64
saving 171
Energy Saving Trust 171
English Nature 81
EPICA *see* European Project for Ice Coring in Antarctica
equinoxes 15-16
ethanol 157
Ethiopia: flooding 8
Europe:
  agriculture 119
  biofuels 158-9
  ice cover 14
  sea levels 115
European Project for Ice Coring in Antarctica (EPICA) 30, 97
European Union: greenhouse gas emissions 145-6
ExxonMobil 50

factories: effects of 8, 9, 21
farming *see* agriculture
Ferrel cell 55
fish 81-2, 116, 121
flexibility mechanisms 130, 134-6
flooding 8, 33, 35, 71-2, 129
food:
  carbon footprint 181-3
  shortages 34
food chain 29
foraminifera *16*, 17, 98
fossil fuels 45, 48-9, 64, 129, 142-3
  replacing 150
Foul Bay, Australia 113
Francis, Justin 180
frogs 84, 86-90

G8 summits 142
Gaia Hypothesis 91-3
Gambia 71
gasification 164
general precession 16
geology: terrestrial: studying 17-18
geothermal energy *161*, 162-3
Gibraltar 29
GISP2 project 24
glacial sediments 14
glaciers 29, 31, 32, 33, 47
  mountain 77-80
  retreating 74-5, 78-9
  types 73
Global Carbon Project 128
Global Change Biology 88
Global Commons Institute 143
global dimming 45
Global Species Assessment

122
global warming:
  economics 33
  human effects on 46-50
  measurements quantifying 48
  rise 8
global warming potential (GWP) 48
Golden Toads 84, 87
Golding, William 91
gravitation 15-16
Gravity Recovery and Climate Experiment (GRACE) 75
Great Barrier Reef 102
greenhouse effect 8-9, 21, 38, *39*
  runaway 99
greenhouse gases 9-10, 23
  absorption 38
  concentrations of 96-7, 110
  by countries 133
  cuts in 10
  dangerous levels 140-4
  emissions:
    by sources *131*
    cuts required 144
    global budget 143-4
    setting ceiling 143-4
    spreading load *143*
    targets 129-36, 140-5
  prediction scenarios 104-5
  reducing levels 94
  warming agents 39-44
  world's biggest emitters 145-8
Greenland:
  earthquakes 76
  ice cap 24-5
  ice sheet 28, 30, 31, 32, 73, 95
  rocks from 25
  temperatures 24, 47
GRIP project 24
Grissino-Mayer, Henri 101
Gulf Stream 25, 30, 57, 58
GWP (global warming potential) 48

Hadley cell 55, 68, 72
halocarbons 42, 43-4
halogens 43
Hansen, Jim 32, 108, 109, 141
Hays, Jim 18
HCFCs 44, 129
health 120-1, 125, 126, 127
heat: distribution 53-5
heat waves 33, 35, 73, 120, 129
Heinrich, Hartmut 25
Heinrich events 25, 26, 30

Himalayas 80
Hitchcock, Dian 92
Hoffert, Marty 144
holidays 122-4
Holocene 24
hominin 70
human activities: affecting climate 8, 31, 46-50, 52-93
hurricanes 33, 34, *52*, 55-7, 68-9, 100-1
hydrofluorocarbons (HFCs) 44, 129
hydrogen: power from 163-4
hydropower 159-61

ice:
  analysis of cores 22-3, 24-5, 49, 95-7, *98*
  pressure exerted 76
ice ages 13-14, 17, 18, 24, 47, 95
  astronomical changes and 14
  causes of 20-1, 22
  cycles 17, 18-19, *19*
  forcing 19-21
  temperature changes in 24
ice caps:
  analysis 22-3, 24-5, 95
  CO_2 bubbles in 22
ice sheets 28, 73, 74, 125, 126, 127
icebergs 25
Iceland 25
  carbon emission reductions 152
Imbrie, John 18
Impacts of Climate Change on Wildlife 81
India:
  carbon emission reductions 143
  flooding 8
  glaciers 79
  greenhouse gas emissions 148-9
  monsoon 29, 97
individuals: taking responsibility 165-83
industrialization 21
infrared radiation 21, 38
insects: disease-carrying 121
Inter-Tropical Convergence Zone (ICTZ) 54, 55
interglacials 17, 18-19, *19*, 24, 95
Intergovernmental Negotiating Committee (INC) 137
Intergovernmental Panel on Climate Change *see* IPCC

International Conference on Climate Change and Tourism 122-3
International Energy Agency 142, 150
ionosphere 36
IPCC (Intergovernmental Panel on Climate Change) 4, 8, 31, 110, 113, 137
reports 31-2, 46, 47, 57-8, 101, 104, 106, 109, 112-13, 138, 139
irrigation 24
islands:
  agriculture 119-20
  sea levels 114
  tourism 123
isotopes 17-18, 96, 98

Japan: greenhouse gas emissions 148
jet streams 55, 61, 72
Joint Implementation 134
Joint Nature Conservation Committee 89

Kaser, Georg 80
Keeling, Charles 21
Kenya: carbon emission reductions 143
Kepler, Johannes 14
Kilimanjaro, Mount 77, 79, 80
King, Sir David 149, 151
Knies, Dr Gerhard 159
krill 83
Kutzbach, John 23
Kyoto Protocol 10, 104, 129-30, 133, 134, 139, 143, 178

La Niña 61, 67-8, 86, 119
Laboratory for Glaciology and Geophysics 23
landslides 65-7
Laurance, Bill 66
leisure: green 180-1
locusts 71, 72
logging 69
Lohachara 8
Lovelock, James 91-2, 93, 93
low-carbon world 149

McKinsey Global Institute 150
Manabe, Syukuro 'Suki' 108
Mann, Michael 49
Margulis, Lynn 91
Marine Climate Change Impacts Partnership 81, 82
marine life 81-3, 92, 116
Marsham, Robert 88

Mauna Loa Volcano 21
Mayans 99
meltwater 27, 74
Mencken, J.L. 176
Meridional Overturning Circulation (MOC) 25-7, 28-30, 57-60, 63, 71, 108
mesosphere 36
methane 29, 41-2, 96, 97, 98, 178
  concentration in atmosphere 42
  emissions 129
  generating water vapour 44
  GWP 48
  in ice 23-4
  increases in 23-4
  production 23
Milankovitch, Milutin 14, 17, 18, 19, 23
Milankovitch cycles 47, 64, 70
Mintz, Yale 108
MOC see Meridional Overturning Circulation
Molnia, Bruce 76
Monbiot, George 50, 180
monsoons 29, 35, 54, 62-3, 70-1, 97
Monteverde 84, 87
Montreal Protocol 29, 43
moon 38
Mora, Professor Claudia 101
mountains 123
  glaciers 77-80

NASA (National Aeronautics and Space Administration) 12
  Godard Institute for Space Studies 13, 23, 32, 108
National Center for Atmospheric Research (NCAR) 45, 56, 61, 108, 109
National Oceancographic Centre, Southampton (NOCS) 57
National Oceanic and Atmospheric Administration (NOAA) 12
National Research Council (NRC) 50
Nature 30
Nepstad, Daniel 69
Netherlands 13, 32, 111
New Scientist 133
New Zealand: agriculture 117-19
Newton, Isaac 14
Niger: flooding 8

Nisbet, Euan 133
nitric acid 178
nitrogen 20, 67
nitrogen dioxide 178
nitrous oxide 43, 48, 96, 129
non-carbon-based economy 166-70
North America:
  agriculture 119
  ice cover 14
  sea levels 114
North Atlantic Deepwater Formation (NADW) 26, 57, 58
North Atlantic Drift 25-6, 27-30, 28, 30, 57
North Atlantic Oscillation (NAO) 60, 62-3, 81
North Greenland Ice Core Project (NGRIP) 31
Norway 76-7, 80
Norwegian Institute for Air Research 44
Norwich Conference 81
Nott, Jon 102
nuclear bombs 97
nuclear energy 149, 150-1, 166

oceans:
  absorbing CO_2 63-4
  calcium carbonate in 22
  climate cycles 60-3
  currents 24, 25-30, 57-60, 59
    vertical 28, 57
  effect on climate 57-61
  food supply 29
  marine algae in 92
  pH levels 63-4
  temperatures 64
  winds 24, 57
Oerlesmans, Hans 80
oxygen 17-18, 96
  absorbing heat 20
ozone 36, 38, 42, 42-3, 178
  absorbing heat 20
ozone layer: hole in 29, 30, 32, 43

Pacala, Stephen 144
Pacific Ocean 99
  temperatures 60-1
Palaeocene/Eocene Thermal Maximum 98
Paris: 'beach' 12
Parker, David 47
peat 14
penguins 83-6, 86
perfluorcarbons (PFCs) 129
perfluoromethane 44
permafrost 29
phenology 88-9
Philip Morris 50

Phillips, Norman 108
phosphorus 67
photosynthesis 39, 41, 66, 67, 69
photovoltaic cells 159
phytoplankton 116
plankton 29, 63-4, 81, 82, 98
  CO_2-absorbing 163, 164
plants:
  drugs from 122
  effects of climate change on 88-9, 121-2
Plass, Gilbert 21
Pliocene 141
polar bears 84, 90-1, 90
Polar cell 55
polar regions 73-6
Potsdam Institute for Climate Impact Research 71, 112
precession of equinoxes 15, 17
predictions 94-127
  computer assistance 105-6, 107-9

radiation:
  electromagnetic 38
  infrared 21, 38
  solar 17, 18, 19, 23, 36, 38, 45, 52-3, 100
radiative forcing 46
radiocarbon dating 18
rainforests 28, 30, 64-70, 65
  drought in 68, 69
  logging 69
ratio spectrophotometer 20
Reid, John 118
respiration 39-40
Roaring Forties 55
rocks:
  breaking down 17
  extracting cores from 17
  sedimentary layers 25, 97-9
  sequences: dating 17-18
Roman Empire 100
Royal Meteorological Society 88
Royal Society for the Protection of Birds 81, 155
runaway greenhouse effect 99
Russia: greenhouse gas emissions 147-8

Saabye, Hans Egede 60
Sahara Desert 29, 30, 31, 70-2, 129
salinity valves 29
Sauber, Jeanne 76
sceptics 46-50
Schellnhuber, Professor

John 28, 30-1
sea level:
    countries below 32
    elements contributing to
        112
    rises in 8, *9*, 31-2, 35, 76-
        7, 80, 141
        impacts of 114-15
        predictions 111-13,
            125, 126, 127
seabirds 86-7
seasons 15, 16
sediment cores 25, 97-9
Selley, Professor 105
Shackleton, Nick 18
sharks 82
shells 64, 92
ships' logs 102
Siberia: permafrost 29
Singapore 115
Sir Alister Hardy Foundation
    for Ocean Science (SAH-
    FOS) 81
skiing 122, 123-4
Smagorinsky, Joseph 108
snow 22, 53
snowfalls 34, 35
Socolow, Robert 144
sodium 24, 97
solar activity 47-8
solar cycles 47
solar energy *158*, 159
solar radiation 17, 18, 19,
    23, 36, 38, 45, 52-3, 100
South America: agriculture
    119
Speedie, Colin 82
sporting events 181
Standard Mean Ocean
    Water 96
Stern, Sir Nicholas 33, *33*,
    124
Stockholm 166-8
stratosphere 36, 38, 42, 45,
    178
subtropics: shift in edge of
    72
sulphates 45

sulphur dioxide 45
sulphur hexafluoride 44,
    129
sulphur oxides 178
sun:
    see also solar ...
    reflecting rays from 164
sunlight 38
Svalbard 25
Sweden: carbon emission
    reductions 143
Switzerland: glaciers 79

temperature:
    changes 24, 25, 38, 72
    elements contributing to
        46
    increases in 9-10, 13,
        106, 110, 129
        changes predicted
            125-7
        dangerous 140-1
    lowest 73
    monitoring 12-13
    record of 49
    warmest years 13
thermohaline circulation
    (THC) 57
thermosphere 36-7
Three Gorges Dam 160-1,
    160
Tibetan Plateau 29, 30, 31
tidal power 161-2
tilt 15, 17
Timbukto 71
timelines 184-91
tipping points 28-9, 30
toads 84, 87
tourism 122-4
Townsend, Alan 67
trade winds 53-4, *62*, 67
transport 132-3, *132, 142*
    reducing use of 174-80
    using biofuels 157-8
tree rings 49, 88-101, *100*
trees 88-9
    offsetting CO2 176-7
Trieb, Dr Franz 159

tropical rainforests *see* rain-
    forests
tropical storms 34
tropics: width 72
troposphere 36, 42, 76, 178
tsunamis 77
turtles 85
Tyndall, John 19-20
Tyndall Centre for Climate
    Change Research 178
typhoons 35, 55

ultraviolet light 36, 38
United Kingdom:
    carbon emission reduc-
        tions 152
    CO$_2$ cuts 10
    sea levels 114
    temperatures 13
United Nations:
    Environment Programme
        (UNEP) 31
    Framework Convention
        on Climate Change
        (UNFCCC) 10, 104, 129-
        31, 137, 140
University of East Anglia
    (UEA) 12
uranium 150-1
US Geological Survey 74
United States of America:
    flooding 8
    greenhouse gas emis-
        sions 10, 144

vegetable oil methyl esters
    (VOME) 157
Venice 102-5, *103*
Vietnam 115
vines: wild 104-5
Visigoths 100
volcanic eruptions 45, 47,
    76, 97
VOME (vegetable oil methyl
    esters) 157
Vostok 23, 73

Walker circulation 61

warming agents 39-44
water: fresh 118, 125, 126,
    127
water vapour 20, 44, 110,
    178
wave power 161-2
weather: extreme 32-3, 125,
    126, 127
West Antarctic Ice Sheeet
    28, 31
wetlands 24
whales 82-3
White, Gilbert 88
Wigley, Tom 45
wildlife:
    effects of climate change
        on 80-91, 121-2
    extinction 33, 64
    impact of wind turbines
        on *154*, 155
wind energy *128*, 151-4,
    154-5, *154*, 166
wind systems 53-7, *53*
winter sports 122, 123-4
Wittman, Dr Hannah 177
Woodland Trust 88
World Conservation
    Monitoring Centre 81
World Conservation Union
    90-1
World Energy Council 150
World Glacier Monitoring
    Service 77
World Health Organization
    120
World Meteorological
    Organization (WMO) 31,
    56, 133
    Global Atmosphere
        Watch (WMO-GAW) 8
Worldwatch Institute 159
WWF 81, 169

Younger Dryas event 26-7

Acknowledgements

The author would like to thank the following people for providing contacts and information sources, and reading draft chapters during the early preparation of this book: Dr Eric Wolff (British Antarctic Survey), Professor Michael Schlesinger (Department of Atmospheric Sciences, University of Illinois), Dr David Vaughan (British Antarctic Survey), Dr Valborg Byfield (National Oceanography Centre, Southampton), Dr Peter Challenor (National Oceanography Centre, Southampton), Keith Fenwick (Met Office), Wayne Elliott (Met Office), Roger Dettmer (formerly Institution of Engineering and Technology), Dr Ted Nield (Geological Society of London), Louise Murray, Roger and Jane Fry, Jacqueline and Trevor Deffee, Alex Benwell, plus Angela and David Benwell.